THE TRANSCENDENTALS AND THEIR FUNCTION IN THE METAPHYSICS OF DUNS SCOTUS

Franciscan Institute Publications

PHILOSOPHY SERIES NO. 3

Edited by Philotheus H. Boehner, O. F. M. *and* Allan B. Wolter, O. F. M.

THE TRANSCENDENTALS AND THEIR FUNCTION IN THE METAPHYSICS OF DUNS SCOTUS

By ALLAN B. WOLTER, O. F. M., Ph. D.

Published by
THE FRANCISCAN INSTITUTE
ST. BONAVENTURE, N. Y.
1946

Nihil Obstat:
JOHN K. RYAN, Ph.D.
Censor Deputatus.
April 30, 1946.

Imprimi Potest:
WENCESLAUS KRZYCKI, O.F.M.
Minister Provincialis.
May 5, 1946.

Imprimatur:
✠ MICHAEL JOSEPH CURLEY, D.D.
Archiepiscopus Baltimorensis et Washingtonii.
May 1, 1946.

THE WICKERSHAM PRINTING COMPANY
LANCASTER, PENNSYLVANIA

To

Father Philotheus Boehner, O.F.M., Ph.D.

in token of my sincere esteem and gratitude

TABLE OF CONTENTS

CAN THERE BE FOCUS?

IS FOCUS POSS?

WHAT OF THE ELLIPSE W/ 2 FOCI?

INTRODUCTION

Within the past decade and a half a marked revival of interest in the thought of Duns Scotus has become apparent. Yet there is one important phase of his thought that remains virtually untouched, his theory of transcendentality. This is strange in view of the fact that Scotus defines metaphysics as the " science of the transcendentals." Speaking of the *materia circa quam*, which, he says, is called by some the subject, or more properly the object of a science, he writes: " As to this object, it has been previously shown that this science has to do with the transcendentals."[1] A knowledge of his theory of the transcendentals therefore is indispensable in understanding Scotus' ideal of a systematic and scientific metaphysics. The present study attempts to satisfy this need by providing a coherent and fairly comprehensive picture of his conception of transcendentality and of the function of each of the more important classes of transcendentals.

Schopenhauer remarks in his *Kritik der kantischen Philosophie* that it is much easier to point out the errors and mistakes in the work of a great mind than to give a clear and complete exposition of its worth. While there are several counts on which Scotus' theory might be justly criticized, we have in the main refrained from so doing, preferring a more benign and consistent interpretation wherever there is any reasonable foundation in his statements for such an interpretation. This is true particularly of certain problems in connection with the formal distinction, the theory of univocation and the virtual primacy of being. Many of the deficiencies of his work, after all, can be condoned in view of the tremendous task which Scotus accomplished in clarifying the real issues at stake in so many vital philosophical problems and in subjecting so much of the vague metaphysical notions, the heritage of the idealistic realism of Plato, to rigorous Aris-

[1] *Metaph.* prologus, n. 10; VII, 7a: De isto autem objecto hujus scientiae, ostensum est prius, quod haec scientia est circa transcendentia.

totelian logic. The extent of this contribution can only be realized by those who have taken pains to master the intricacies and technical language of medieval logic, without a knowledge of which Scotus cannot hope to be understood.

Since the time of Christian Wolff, philosophy in general and metaphysics in particular have undergone a process of atomization, with the result that the average student of philosophy has lost all sense of the unity of his science. For one who has glimpsed the ideal of a strict science, as Aristotle conceived and developed it in the first book of the *Analytica Posteriora*, the pseudo-science we have come to know under the pretentious name of ontology is a sorry substitute. As someone once described it, it is little more than a " non-alphabetical dictionary of scholastic terms." While many contemporary neo-scholastics have realized this deficiency and have sought to reduce metaphysics to an orderly and systematic science, with few notable exceptions their efforts have not been too successful.[2] If we have understood Scotus correctly, we believe that his theory of transcendentality offers great possibilities of integrating the science of metaphysics in terms of a theologic [3]—an ideal visualized by Aristotle and St. Thomas alike.

---

[2] Pacificus Borgmann, O.F.M., has done pioneer work in this regard. Unfortunately he has never published in a systematic work his scientific metaphysics, though he has given us numerous articles calling attention to serious defects of the usual neo-scholastic presentation of the " first philosophy " and has sketched the general outlines of an ideal metaphysics which is both modern and scholastic in spirit. The reader is referred especially to his " Gegenstand, Erfahrungsgrundlage und Methode der Metaphysik " in *Franziskanische Studien,* XXI (1934), 80-103; 125-150; " Seiender oder werdender Gott? Substanzialität oder Aktualität des Urseienden? " in *Theologische Gegenwartsfragen* (herausgeg. von E. Schlund, Regensburg, 1940, J. Habbel), pp. 63-81; " Kausaler oder Substanzialer Gottesbeweis? " in *Zeitschrift für den katholischen Religionsunterricht an höheren Lehranstalten,* XIV (1937), 181-195; " Der unvollendete Zustand der aristotelisch-scholastischen Metaphysik " in *Franzisk. Stud.,* XXIII (1936), 404-425.

[3] Borgmann has reintroduced this truly Aristotelian term of theologic or θεολογική (confer Aristotle's *Metaph.* VI 1, 1026a 19) to indicate a scientific metaphysics in which natural theology is not divorced or isolated as a special science but is the predominant and unifying element. As Father

Intriguing as these possibilities may be, we have restricted ourselves in the main to an exposition of the theory of transcendentality as Scotus conceived it, and to emphasizing particularly its theologic implications. While this is not a comparative study, it is interesting to note that if Scotus' theory of the transcendentals represents a unique development in medieval thought, it is at the same time a development essentially in harmony with the best scholastic tradition.

Incidentally, it may be remarked in this connection that the chasm between the Angelic Doctor and the Subtle Doctor is not so unbridgeable as is commonly believed. The author is inclined to suspect that in the majority of instances Scotus does not differ so much from St. Thomas as he does from certain strains of Thomism. If the difference of terminology be taken into account, the doctrines of the two scholastics will, we believe, be found to be complementary more often than contradictory. This is true even of the doctrine of univocation, so fundamental to Scotus' metaphysics. As Scotus himself remarked, the great lights among the schoolmen in their treatises on God and what we can know about Him actually make use of univocation, though they do not admit it in so many words.[4]

This itself, however, is a controversial point. To defend it adequately would require more time and study than we care to devote to it here, and, as we have learned from experience, would only lay us open to the criticism of having substituted one interpretation of Duns Scotus or St. Thomas for another. We have in general, then, refrained from attempting any positive reconciliation and we leave to the discerning reader to discover such possibilities of rapprochement for himself.

Borgmann remarks, we are in a position akin to that of Aristotle. To distinguish his own rational and purely philosophical approach from the theological and cosmogonical speculations of theology of the poets, he adopted the term "theologic." Confer "Gegenstand, Erfahrungsgrundlage und Methode der Metaphysik", *Franz. Stud.*, XXI (1934), 88 (note 11).

[4] *Report. Par.* 1, d. 3, q. 1, n. 7; XXII, 95b: Hoc etiam Magistri tractantes de Deo, et de his, quae cognoscuntur de Deo, observant univocationem entis in modo dicendi, licet voce hoc negent.

In this study only those works recognized as definitely authentic by the *Commissio Scotistica* have been used. A relatively complete bibliography of the editions of the works of Duns Scotus was published by this commission, *Lineamenta Bibliographiae Scotisticae,* by Uriel Smeets, O.F.M. (Rome, 1942, editio pro manuscripto). A list of the authentic works can be found in the article by the Rev. Dr. Maurice Grajewski, O.F.M., "Duns Scotus in the Light of Modern Research" published in the *Proceedings of the American Catholic Philosophical Association,* XVIII (1942), 168-185, or in the report of a conference given by Father Longpré, "Stand der Skotus-Forschung 1933" by the Rev. Dr. Marianus Mueller, O.F.M., in *Wissenschaft und Weisheit,* I (1934), 63-71; 147-153. The reader is also referred to the recent article by the Rev. Dr. Carl Balić, O.F.M., president of the *Commissio Scotistica,* "De Critica Textuali" in *Antonianum,* XX (1945), 267-308.

The references in the footnotes are made to the Vivès edition of the *Opera Omnia* of Duns Scotus with the exception of those pertaining to the *Tractatus De Primo Principio.* The recent critical edition of M. Mueller (Freiburg im Breisgau, Herder, 1941) has been used for the latter. It should be noted, however, that the text of the Vivès edition has been frequently corrected in accord with the more reliable editions mentioned in the bibliography. The bibliography of general works includes merely those works cited in the text. It goes without saying that the fact that a work appears in the general bibliography does not imply that the author endorses its interpretation of Scotus or Scotistic doctrine. For a critically annotated bibliography of the more important recent works on Scotistic thought, the reader is referred to the "Scotistic Bibliography of the Last Decade (1929-1939)" by Father Grajewski, published successively in *Franciscan Studies,* XXII-XXIII (1942-1943), and to Father Bettoni's *Vent' Anni di Studi Scotisti: 1920-1940,* (Milan, 1943).

The writer wishes to take this occasion to thank all who have helped him in preparing this dissertation. In particular the author owes much to his major professor, the Rev. Dr. John K. Ryan, who encouraged to the utmost his interest in the philosophy of Duns Scotus, to the Rev. Dr. Ignatius Smith, O.P., dean of the School of Philosophy, the Rev. Dr. William J. McDonald and the Rev. Dr. Jules Baisnée, S.S., for their critical reading of this manuscript and their helpful suggestions. But above all we are indebted to the Rev. Dr. Philotheus Boehner, O.F.M., director of Studies and Research in the Franciscan Institute, St. Bonaventure, N. Y., for calling our attention to the need of such a study, for the constant scholarly assistance he has rendered the writer and for his critical reading of this manuscript. We also express our thanks to him for the use of his revised texts of the works of William Ockham and for his lecture notes on the *History of Franciscan Philosophy.* To the Rev. Dr. Carl Balić, O.F.M., president of the *Commissio Scotistica,* we are grateful for information regarding the authenticity of certain writings attributed to Scotus; to the Rev. Joachim Daleiden, O.F.M., M.A., for obtaining indispensable manuscripts and editions of various works of Scotus and Scotistic commentators, to various libraries and librarians, particularly to the Rev. Irenaeus Herscher, O.F.M., M.A., B.L.S. of the Friedsham Memorial Library, St. Bonaventure's College, and Brother Henry Demko, O.F.M., librarian of the Franciscan Monastery, Washington, D. C.

This dissertation was submitted to the Faculty of the School of Philosophy of the Catholic University of America in partial fulfillment of the requirements for the degree of Doctor of Philosophy.

List of Principal Abbreviations used for Scotus' Works

Oxon. 1, d. 3, q. 3, nn. 18-19; IX, 597b-598a. *Opus Oxoniense,* book I, distinction 3, question 3, marginal numbers 18 to 19; volume IX (Vivès edition), pages 597 (right column) to 598 (left column).

Rep. Par. 2, d. 34, q. un., n. 3; XXIII, 170a-b. *Reportata Parisiensia,* book II, distinction 34, the only question, marginal number 3, volume XXIII, page 170 (both columns).

Metaph. 6, q. 4, n. 3; VII, 349a. *Quaestiones subtilissimae super libros Metaphysicorum Aristotelis,* book 6, question 4, marginal number 3, volume VII, rest read as above.

Coll. 13, n. 4; V, 202a. *Collationes seu disputationes subtilissimae*, collatio 13, rest read as above.

Quodl. q. 3, n. 2; XXV, 114a. *Quaestiones Quodlibetales*, question 3, rest read as above.

Super univers. q. 4, n. 2; I, 96b. *Super Universalia Porphyrii Quaestiones Acutissimae*, question 4, rest read as above.

Super lib. elench., q. 13, n. 5, II, 17b. *In libros Elenchorum Quaestiones*, rest read as above.

Super Praedicamenta, q. 11, n. 1, I, 465b. *In librum Praedicamentorum Quaestiones*, rest read as above.

De Primo, c. 2, concl. 16, p. 33. *Joannis Duns Scoti Tractatus De Primo Principio* (M. Mueller edition), chapter 2, conclusion 16, page 33.

PART ONE

ON THE NATURE OF THE TRANSCENDENTALS IN GENERAL

The terms "transcendent" and "transcendental" have a variety of meaning even in the realm of philosophy. With the mystically minded "transcendent" connotes something intangible and inaccessible. Schelling identified it with Mind as distinct from Nature. With Kant "transcendent" refers to whatever lies beyond the realm of experience and valid knowledge, while "transcendental" designates the *a priori* and necessary factors in experience. With the scholastics, however, the term has a much more objective and realistic meaning, at least when applied to metaphysical notions as contrasted with the transcendental notions of logic. For the schoolmen understood by "transcendentals" those abstract yet very real concepts which escape classification in the Aristotelian categories by reason of their greater extension and universality of application.

The precise origin of this specific use of the term is shrouded in obscurity. Prantl attributed it to the unknown author of the *De natura generis*, who refers to the six notions of *ens, res, aliquid, unum, verum* and *bonum* as "transcendentia".[1] But since this opusculum, according to Prantl, was probably written after the time of Scotus, Schuleman credited Scotus with the introduction of the term.[2] However, as Knittermeyer had previously indicated,[3] St. Albert the Great already made use of the

---

[1] Carl Prantl, *Geschichte der Logik im Abendlande* (Leipzig, Gustav Fock, 1927), Vol. III, p. 245.

[2] Günther Schuleman, *Die Lehre von den Transcendentalien in der scholastischen Philosophie* ("Forschungen zur Geschichte der Philosophie und der Pädagogik," herausg. von Artur Schneider und Wilhelm Kahl, Bd. IV, hft. 2) Leipzig, Felix Meiner, 1929, p. 46.

[3] Hinrich Knittermeyer, *Der Terminus transszendental in seiner historischen Entwickelung bis zu Kant* (Marburg, J. Hamel, 1920) pp. 16-17.

term in this sense. Even earlier, Roland of Cremona applied it to *ens, unum, aliquid* and *res.*[5]

Whether or not Roland is to be regarded as the author of this terminology, it is certain that the idea of "transcendentality," in the sense of transcending the categories or ultimate genera, is of much earlier origin. Whatever is to be said of the Pythagorean theory of number, or of Anaxagoras' Νοῦς, or of Plato's Ἰδέα as the first glimmerings of a transcendental philosophy,[6] we have in Aristotle a very definite idea of transcendentality as Scotus was later to define it.[7]

It is not our purpose, however, to sketch the history of the transcendentals as such; to indicate, for instance, the use of the scholastic formula, "*Unum, verum et bonum convertuntur cum ente*" by Alexander of Hales,[8] the classical analysis of the co-extensive transcendentals by St. Thomas,[9] or the recognition of the implications of the disjunctive transcendentals as a distinct group by St. Bonaventure,[10] and so on. Such a task falls to works of a more general nature than ours.[11]

---

[4] *Liber de praedicabilibus,* tract. 4, c. 3; *Opera Omnia,* ed. A. Borgnet (Paris, Vivès, 1890-1899), I, p. 64b; *Metaphysicorum Libri tredecim,* lib. I, tract. 1, c. 2; *Opera,* VI, p. 6a.

[5] *Summa theologica,* Paris, Mazarine Mss. 795, fol. 7vb: Nisi esset unum de transcendentibus, scilicet ens, unum, aliquid, res. (quoted from H. Pouillon, "Le premier traité des propriétés transcendentales" in *Revue Néo-scolastique de Philosophie,* XLII (1939) 44.)

[6] Schulemann, *op. cit.,* pp. 1-4.

[7] Confer Aristotle's discussion of *being* and *unity* as transcending all genera. *Metaphysics,* III, 3, (*Opera Omnia,* ed. Bekker. Berlin, 1831-1870), 998b 20 ff.

[8] *Summa theologica,* P. I, inq. 1, tract. 3, qq. 1-3, passim (Quaracchi, Collegium S. Bonaventurae, 1924-1930), I, pp. 112-200.

[9] *Quaestio disputata de Veritate,* quaest. 1, art. 1.

[10] *Quaestio disputata de Mysterio Trinitatis,* q. 1, art. 1; *Opera Omnia* (Quaracchi, Coll. S. Bonaventurae, 1882-1902), V, pp. 46-47.

[11] For a general survey confer such works as that of Schulemann, *op. cit.,* Knittermeyer, *op. cit.,* Henri Pouillon, "Le premier traité des propriétés transcendentales. La *Summa de bono* du Chancelier Philippe" in *Revue Néoscolastique de Philosophie,* XLII (1939), 40-77; Johann Fuchs, *Die*

This study of Scotus' notion of the transcendentals falls into two main sections. The first deals with the idea of transcendentality in general and its basic doctrinal presuppositions; the second discusses the various classes of transcendentals in particular. A separate chapter is appended by way of conclusion, indicating how Scotus welded the heterogeneous elements of Aristotle's unfinished metaphysics into a single organic whole as the "science of the transcendentals."

Proprietäten des Seins bei Alexander von Hales (München, Druck der Salesianischen Offizin, 1930); Heinrich Kühle, "Die Lehre Alberts des Grossen von den Transzendentalien" in *Philosophia Perennis* (Regensburg, 1930) I, 129-157.

CHAPTER I

Definition and Division of the Transcendentals

It is important to emphasize at the outset that this is a study of the transcendentals in so far as they pertain to metaphysics and consequently have reference to the order of reality and existence. For logic too has its transcendental notions,[12] notions which deal with pure thought forms and their interrelations and in that sense transcend the order of existence itself. Contrary to what is sometimes believed, Scotus carefully distinguishes between the two orders.[13] While his speculation on the interrelation of the two spheres of transcendentality would make an interesting study, it lies beyond the scope of this work.

Rather than attempt to project a ready-made definition into the texts of Scotus, we shall let him speak for himself. Fortunately, his Oxford *Commentary on the Sentences of Peter Lombard* contains a passage on the subject of transcendentals, the most extensive treatment found in any one place of his authentic works. Because of its illuminating character it may be regarded as a key-text and is used as the basis of this study.

> Now a doubt arises as to what kind of predicates are those which are predicated formally of God, for instance, *wise, good,* etc. I answer, that before being is divided into the ten categories, it is divided into infinite and finite. For the latter, namely finite being, is common to the ten *genera*. Whatever pertains to being then in so far as it remains indifferent to finite and infinite, or as proper to the Infinite Being, does not belong to it as determined to a genus, but

[12] Confer, for instance, Scotus' *Quaestiones subt. super libros Metaphysicorum Aristotelis,* lib. 6, q. 3, n. 15 (Wadding edition reedited by Vivès, Paris, 1891-1895). VII, 346. *Super Universalia Porphyrii,* q. 27, n. 7; I, 316a; etc.

[13] *Metaph.* 6, q. 1, tota quaestio, VII, 302f; *Super Praedicamenta,* q. 1, n. 4; I, 438b, etc.

prior to any such determination, and therefore, as transcendent and outside of any genus. Whatever [predicates] are common to God and creatures are of such kind, pertaining as they do to being in its indifference to what is infinite and finite. For in so far as they pertain to God they are infinite, whereas in so far as they belong to creatures they are finite. They belong to being, then, prior to the division into the ten genera. Anything of this kind, consequently, is transcendent.

But then another doubt arises. How can *wisdom* be considered a transcendental if it is not common to all beings, for transcendentals seem to be common to all? I answer that just as it is not of the nature of a supreme genus to have several species contained under it, but rather not to have any genus over and above it (the category *quando,* for instance, is a supreme genus since it has no genus over and above it, although it has few, if any, species contained under it), so also whatever is not contained under any genus is transcendental. Hence, not to have any predicate above it except being, pertains to the very notion of a transcendental. That it be common to many however is purely incidental. This is evident too from the fact that *being* possesses not only attributes which are coextensive with it, such as *one, true* and *good,* but also attributes which are opposed to one another such as *possible-or-necessary, act-or-potency,* and such like.

But if the coextensive attributes are transcendental because they pertain to being as undetermined to a definite genus, then the disjunctive attributes are transcendental too. And both members of the disjunction are transcendental since neither determines its determinable element to a definite genus. Nevertheless one member of the disjunction is proper and pertains formally to one being alone, for instance, *necessary* in the disjunction *necessary-or-possible,* or *finite* in the disjunction *finite-or-infinite,* and so also with the others.

And so *wisdom* or anything else, for that matter, which is common to God and creatures can be transcendental. A transcendental, however, may also be predicated of God alone, or again it may be predicated about God and some creature. It is not necessary then that a transcendental *as transcendental* be predicated of *every* being, unless it be coextensive with the first of the transcendentals, namely *being.*[14]

---

[14] *Opus Oxoniense,* lib. 1, dist. 8, quaest. 3 nn. 18-19; IX, 597b-598b: Sed tunc est dubium, qualia sunt illa praedicata, quae dicuntur de Deo [formaliter], ut *sapiens, bonus,* etc.—Respondeo: ens prius dividitur in

This text, found in the Vivès edition but with minor corrections based on seventeen of the best manuscripts, belongs to the *Ordinatio Scoti* and does not represent an addition made from the *Additiones Magnae* or other *Reportata*.[15] It may be safely used

infinitum et finitum quam in decem genera [praedicamenta *for* genera A], quia alterum istorum, scilicet (ens) finitum, est commune ad decem Genera. Ergo quaecumque conveniunt enti ut indifferens ad finitum, et infinitum, vel ut est proprium enti infinito, conveniunt sibi non ut determinatur ad genus sed ut prius, et per consequens, ut est transcendens et est extra omne genus. Quaecumque sunt communia Deo et creaturae, sunt talia quae conveniunt enti ut est indifferens ad finitum et infinitum; ut enim conveniunt Deo sunt infinita, ut creaturae sunt finita. Ergo illa per prius conveniunt enti quam ens dividatur in decem genera, et per consequens quodcumque tale est transcendens.

Sed tunc est aliud dubium, quomodo ponitur sapientia transcendens, cum non sit communis omnibus entibus, (et transcendentia videntur communia omnibus).—Respondeo, sicut de ratione (generis) generalissimum non est habere sub se plures species, sed non habere aliud supraveniens genus sicut hoc praedicamentum Quando—quia non habet supraveniens genus—est generalissimum, licet paucas habeat species aut nullas, ita transcendens quodcumque nullum habet genus sub quo contineatur. Unde de ratione transcendentis est non habere praedicatum supraveniens nisi ens. Sed quod ipsum sit commune ad multa inferiora, hoc accidit. Hoc patet ex alio: quia ens non tantum habet passiones simplices convertibiles, sicut *unum, verum, et bonum,* sed habet aliquas passiones ubi opposita distinguuntur contra se, sicut *necesse* [*esse*] *vel possibile, actus vel potentia,* et hujusmodi. Sicut autem passiones convertibiles sunt transcendentes, quia consequuntur ens non inquantum determinatur ad aliquod genus, ita passiones disjunctae sunt transcendentes; et utrumque membrum illius disjuncti est transcendens, quia neutrum determinat suum determinabile ad certum genus; et tamen unum membrum illius disjuncti formaliter est speciale non conveniens nisi uni enti, sicut *necesse* in ista divisione *necesse vel possibile,* et similiter *infinitum* in ista divisione *finitum vel infinitum,* et sic de aliis.

Ita etiam potest *sapientia* esse transcendens et quodcumque aliud quod est commune Deo et creaturae, licet aliquod tale dicatur de solo Deo, aliquod autem de Deo et aliqua creatura. Non oportet autem transcendens ut transcendens dici de quocumque ente, nisi sit convertibile cum primo transcendente, scilicet ente. [Note: The above text, while not intended as a critical so much as a "safe" reading, has been corrected on the basis of variants of the following codices, whch have been graciously supplied by Father Balić of the Scotistic Commission: APVBRMNCJSTLQZGΔD. What is omitted by A alone is put in parentheses; what is found in A alone is put in brackets. For the *tabula codicum* see C. Balić, *Ratio Criticae Editionis Operum Omnium J. Duns Scoti,* II (1939-1940), pp. 73-74.]

[15] For a list of the sections in the Oxford Commentary on the Sentences (*Opus Oxoniense*) that do not belong to the original *Ordinatio Scoti*

therefore as the basis for our discussion of Scotus' theory of transcendentals.

A careful analysis of this key-text reveals a number of important points and at the same time outlines the principal problems of our study.

THE TRANSCENDENTAL: A REAL CONCEPT

It is significant that Scotus introduces his tract on the transcendentals from the standpoint of predication. "What kind of predicates are they which are predicated formally of God?"[16] Predication is an intellectual operation and has to do primarily with concepts or notions. It is important to remember that the transcendentals are fundamentally notions or concepts, particularly when discussing the relation of being and its attributes. By forgetting this fact, one may easily read into Scotus contradictions which exist only in the mind of the interpreter.

On the other hand, though the transcendentals are notions, they are very real concepts. That is to say, they do not refer to the conceptual order but to the metaphysical order of reality.[17] They are real in the sense that they stand for and signify immediately an existing entity. Scotus indicates this when he speaks of these notions being "predicated formally of God."[18]

---

but were supplied from other works, principally the *Additiones Magnae* and the *Reportata,* see Carl Balić, *Relatio a Commissione Scotistica exhibita Capitulo Generali Fratrum Minorum Assisii A.D. 1939* (Rome, 1939), p. 141ff.

[16] *Oxon.* 1, d. 8, q. 3, n. 18; IX, 597b: Qualia sunt illa praedicata, quae dicuntur de Deo formaliter, ut *sapiens, bonus,* etc.

[17] Confer Timothy Barth, "De fundamento univocationis apud J. Duns Scotum" in *Antonianum,* XIV (1939), 287ff.

[18] "Formaliter" as opposed to "virtualiter" implies that the *ratio* or perfection *as such* is to be found *intrinsically* in the thing of which it is predicated. Confer Scotus, *Reportata Parisiensia,* lib. 1, dist. 45, quaest. 2, n. 5; XXII, 500b-501a: Expono etiam hoc vocabulum *formaliter.* Dico autem esse formaliter tale, sive esse in alio formaliter quod non est in eo potentialiter, ut album in nigro, nec virtualiter, ut effectus in sua causa est. Nec hoc dico formaliter esse in aliquo quod est in eo confuse et cum quadam commixtione, (quomodo ignis est in carne non formaliter); sed

In the following chapter, the conditions required on the part of the object for the validity of a real notion will be discussed more at length. For the present it suffices to note that the transcendentals are predicated of real things and signify some formal aspect or perfection characteristic of existing objects.

THE DEFINITION OF THE TRANSCENDENTAL

Scotus indicates the essential nature of a transcendental when he says: "*Whatever cannot be contained under any genus is transcendental.*"[19] Or in more positive terms: "Whatever pertains to being, then, insofar as it remains indifferent to finite and infinite, or as proper to the Infinite Being... [belongs to it] as transcendental."[20]

In short, it is not required that a transcendental be coextensive, or, as the scholastics put it, convertible with being.[21] Scotus consequently gives the term "transcendental" a much broader application than does, for example, the author of the *De Natura Generis*, who limits the transcendentals to *unum, verum, bonum, ens, res* and *aliquid.*[22] Even if Scotus was not the first to extend the conception of the *transcendentia* beyond these six, his reference to the objection which he anticipates, *transcendentia viden-*

---

dico esse formaliter in aliquo, in quo manet secundum suam rationem formalem et quidditativam, et esse tale formaliter est includere ipsum secundum suam rationem formalem praecisissime acceptam.

[19] *Oxon.* 1, d. 8, q. 3, n. 19; IX, 598a: Ita transcendens quodcumque nullum habet genus sub quo contineatur.

[20] *Ibid.*, n. 18; 597b-598a: Ergo quaecumque conveniunt enti ut indifferens ad finitum et infinitum, vel ut est proprium enti infinito, conveniunt sibi non ut determinatur ad genus, sed ut prius, et per consequens, ut est transcendens et extra omne genus. *Ibid.* 2, d. 1, q. 4, n. 15; XI, 111a: Quidquid convenit enti inquantum est indifferens ad infinitum et finitum, convenit ei prius quam dividatur in genera, et ita est transcendens.

[21] *Ibid.* 1, d. 8, q. 3, n. 19; IX, 598b: Non oportet ergo transcendens ut transcendens, dici de quocumque ente.

[22] *De Natura Generis*, c. 2 (edited by Michael de Maria, S.J. in *Sancti Thomae Aq. Opuscula Philosophica et Theologica.* Tiferni Tiberini, S. Lapi, 1886) Vol. I, p. 286: "Sunt autem sex transcendentia: videlicet: ens, res, aliquid, unum, verum, bonum: quae re idem sunt, sed ratione distinguuntur."

tur communia omnibus,[23] would seem to indicate that he is at least departing from the current interpretation of the term.

Yet this broader application of the term is not arbitrary but represents on the contrary a return to the position of St. Albert himself.[24] For there are many real notions that do not fit under the categories properly speaking and still are not coextensive with being. All notions that are either common to God and creatures or are proper to God alone transcend the categories.

> And so wisdom, or anything else, for that matter, which is common to God and creatures can be transcendental. A transcendental though may also be predicated of God alone and not of any creature, or again it may be predicated about God and some creature. It is not necessary then that a transcendental as transcendental be predicated of every being.[25]

Scotus, then, defines as transcendental *whatever rises above all genera and transcends all categories.* He draws a comparison. It makes little difference whether a category or supreme genus contains many *species* or none at all. For the essence of the category or "predicament" lies in the fact that it is supreme or ultimate in its own order. It cannot be subordinated to a more universal genus, since no more universal genus exists. In a similar way, the essential note of a transcendental, as its very name indicates, is that it escapes from categorical classification. Whether it be proper to one individual or whether it be coextensive with all reality is purely incidental.[26]

Scotus returns to this basic idea again and again. "Whatever is not contained under any genus is transcendental,"[27] "not...

---

[23] *Oxon.* 1, d. 8, q. 3, n. 18; IX, 598a: Sed tunc est aliud dubium quomodo ponitur *sapientia* transcendens, cum non sit communis omnibus entibus, et transcendentia videntur communia omnibus.

[24] St. Albert enumerates disjunctives among the *passiones entis* in the same passage where he speaks of *transcendentia.* Cf. *Metaphy.* 1, tr. 1, c. 2; VI, 5-6.

[25] *Oxon.* 1, d. 8, q. 3, n. 19; IX, 598b.

[26] *Ibid.,* 598a: Sed quod ipsum sit commune ad multa inferiora, hoc accidit.

[27] *Ibid.*

as determined to a genus, but prior to any such determination, and therefore, as transcendent," [28] "the coextensive attributes are transcendental because they pertain to being as undetermined to a definite genus," [29] " and both members of the disjunction are transcendental since neither determines its determinable element to a definite genus," [30] " They belong to being, then, prior to the division into the ten genera. Anything of this kind consequently is transcendent." [31]

For this reason, " finite " itself is a transcendental notion even though it is neither common to God and creatures nor proper to the Infinite Being.[32] Yet it has its rightful place among the transcendentals since it is not confined to a single genus but is common to all.[33]

DIVISION OF THE TRANSCENDENTALS

Within this broad classification four distinct types or layers of transcendentality that are of special importance in Scotus' theory may be singled out. They are all enumerated in the passage under consideration. In the order of their universality, though not necessarily in the order of their relative importance, they are listed as follows:

*1. Being (*ens*), which Scotus calls the "first of the transcendentals." [34]

---

[28] *Ibid.*, n. 18; 597b-598a.

[29] *Ibid.*, n. 19; 598a.

[30] *Ibid.*

[31] *Ibid.*, n. 18; 598a.

[32] *Ibid.*, n. 19; 598a-b: Et utrumque membrum illius disjuncti est transcendens, quia neutrum determinat suum determinabile ad certum genus; et tamen unum membrum illius disjuncti est speciale formaliter non conveniens nisi uni enti, sicut...*infinitum* in ista divisione *finitum vel infinitum.*

[33] *Ibid.*, n. 18; 597b: Ens finitum est commune ad decem genera.

[34] *Ibid.* 597b-598a: Ens...ut est transcendens et extra omne genus; *ibid.*, n. 19; 598b: Non oportet ergo transcendens ut transcendens, dici de quocumque ente nisi sit convertibile cum *primo transcendente,* scilicet cum *ente.* (Italics mine).

2. The properties or attributes coextensive with being as such (*passiones entis simpliciter convertibiles*), for instance, unity, truth, goodness.[35]
3. The disjunctive attributes, such as "infinite-or-finite," "substance-or-accident," "necessary-or-contingent," and so on. These attributes, which in general are the primary differences of real being, are in disjunction proper to being and in that sense coextensive with real being.[36]
4. Finally, we have all those remaining "pure perfections" which have not been included in one of the above types. The pure perfections (*perfectiones simpliciter*) include being and its coextensive attributes as well as the more perfect member of each disjunction. As a class, they have a special function in the metaphysics of Scotus.[37] The pure perfections are transcendental for the simple reason that they can be predicated of God and hence transcend the finite categories.[38] They are of two kinds:
   a. perfections predicable of God alone, such as omnipotence, omniscience, etc.[39]
   b. perfections predicable of God as well as of certain creatures, such as wisdom, knowledge, free will, etc.[40]

Two other considerations are to be noted in connection with this analysis of the concept of the transcendentals. They arise not so much from the key-text itself as from the context in which

---

[35] *Ibid.*, 598a: Ens...habet passiones convertibiles simplices, sicut *unum, verum, bonum,* etc....Passiones convertibles sunt transcendentes quia consequuntur ens non inquantum determinatur ad aliquod genus.

[36] *Ibid.:* Ens non tantum habet passiones convertibiles simplices...sed habet aliquas passiones ubi opposita distinguuntur contra se, sicut *necesse esse vel possibile, actus vel potentia,* et hujusmodi....Et passiones disjunctae sunt transcendentes, et utrumque membrum illius disjuncti est transcendens.

[37] *Ibid.*, d. 3, q. 3, n. 7; IX, 104a: Omnia transcendentia dicuntur perfectiones simpliciter; *Ibid.* d. 8, q. 3, n. 30; IX, 630a: Omnes sunt transcendentes quae important perfectionem simpliciter.

[38] *Rep. Par.* 1, d. 8, q. 5, n. 13; XXII, 170b: Quidquid enim dicitur de Deo est formaliter transcendens.

[39] *Oxon.* 1, d. 8, q. 3, n. 19; IX, 598b: Aliquod tale [sc. transcendens] dicatur de solo Deo et de nulla creatura."

[40] *Ibid.:* "...aliquod autem de Deo et creatura aliqua..."

it occurs. The first is Scotus' theory of univocation and the other his conception of metaphysics as a theologic.[41]

UNIVOCITY AND THE TRANSCENDENTALS

It is not by chance that Scotus introduces this consideration of the transcendentals after a discussion of the problem of whether or not God falls into a genus?[42] This problem assumes particular difficulties for one who maintains that not merely the notion of being, but all pure perfections that are predicable of God and creatures, are univocal. In fact, the doctrine of univocation runs through the whole theory of the transcendentals. It creates a particular problem in connection with the primacy of being. It differentiates Scotus' notion of transcendental truth and goodness from that of his contemporaries. It is the foundation of all inferences made on the basis of the disjunctive transcendentals. In short, it penetrates the entire theory of transcendentality as he conceived it.

THEOLOGICAL IMPLICATIONS OF TRANSCENDENTALS

The theory of transcendentals is important for Scotus primarily because of its implications for a natural theology or theologic. Not only does the context from which the above passage is taken indicate this, but the opening words of the key-text itself tell us as much. "What kind of predicates are those which are predicated formally of God?" His most extensive

---

[41] Question III in which the key-text occurs reads: "Utrum Deus sit in genere?" The fourth argument for the affirmative opinion is drawn from his previous tract on univocation. It reads: "Item, sapientia formaliter dicitur de Deo, et hoc secundum eamdem rationem, secundum quam dicitur de nobis, quia illae rationes, quae dictae sunt dist. 3. quaest. 2 de univocatione entis, concludunt de univocatione sapientiae; igitur secundum illam rationem secundum quam sapientia dicitur de Deo, est species generis...." *Ibid.* n. 1; IX, 580b.

[42] While Scotus defines transcendentality from a negative viewpoint, when he speaks of it in positive terms, it is always with reference to God. Thus it is either a perfection which is proper to God or one that is common to God and creatures. This merely emphasizes our point in connection with the theologic implications of the transcendentals. Cf. infra.

discussion of transcendentality is, in short, in connection with our natural knowledge of God. And at times Scotus even applies the term " transcendental " exclusively to those perfections which are formally predicable of God—*denominatio a potiori.*[43] In fact the imperfect transcendental notions, like contingent, finite, caused, etc., are important not as an independent class but as members of a disjunction. As such they form the starting point in an inference that ends with God.

The analysis of this fundamental passage both as to text and context reveals the following basic points:

1. A definition, or better, description, of the transcendental in reference to the predicamental genera or categories, which it transcends.
2. Its character as a real notion.
3. Its univocal nature.
4. Its four main divisions or classes: a) being, b) the proper attributes simply coextensive with being, c) the attributes coextensive in disjunction with being, d) the pure perfections.
5. Its theologic implications.

The essential notion or the definition of a transcendental should be sufficiently clear from what has been said so far. The remaining four points require further amplification and will be dealt with in the following chapters in their respective order, a separate chapter being devoted to each of the four divisions of the transcendentals. The bearing of Scotus' theory of transcendentality on a natural theology will become more apparent in the following pages. The principal conclusions in this regard will be correlated in the final chapter.

---

[43] *Quodlibet.* q. 14, n. 3; XXVI, 5-6: Breviter dico, quod quodcumque transcendens per abstractionem a creatura cognita, potest in sua indifferentia intelligi, et tunc concipitur Deus quasi confuse, sicut animali intellecto, homo intelligitur. Sed si tale transcendens in communi intelligitur sub ratione alicujus specialioris perfectionis... jam habetur conceptus sic Deo proprius, quod nulli alii convenit. *Oxon.* 1, d. 3, q. 3, n. 7; IX, 104a: Omnia transcendentia dicuntur perfectiones simpliciter, et conveniunt Deo in summo.

CHAPTER II

THE OBJECTIVE BASIS FOR REAL CONCEPTS

THE transcendentals are real notions or concepts. What, then, is a real concept? A concept is essentially a sign and, like all signs, refers to something other than itself.[1] The *signatum,* or thing referred to, may be either some extra-mental reality (*res extra animam*) or the content of another concept (*res rationis*).[2] In the first case, the concept is real (*conceptus realis*); in the second, it is logical (*conceptus rationis*). Or in the more usual terminology, the former is a " first intention ", since the first activity of the mind tends to know or " intends " a really existing object.[3] The *conceptus rationis* is a " second intention ", since it tends towards or refers to a mental entity, namely, the content of another concept. Since the logical concept presupposes the existence of a real concept or first intention, it is a secondary activity.[4]

From the viewpoint of its origin, a real concept is produced in the possible intellect through the activity of the agent intellect and the phantasm or object contained in the phantasm.[5] The

---

[1] *Oxon.* 4, d. 10, q. 4, n. 3; XVII, 229b: Conceptus sunt signa rerum naturaliter.

[2] *Ibid.* d. 1, q. 2, n. 7; XVI, 107a: Conceptus sit rei extra [animam] sive rationis.

[3] P. Minges, " Der angebliche exzessive Realismus des Duns Scotus " in *Beiträge zur Geschichte der Philosophie des Mittlelalters,* VII (Münster, 1908), 83.

[4] Stefan Swiezaswki, " Les intentions permières et les intentions secondes chez Jean Duns Scot " in *Archives d'Histoire doctrinale et littéraire du Moyen Age,* IX (1934) 205-260.

[5] *Prima lectura,* I, 3, 17, Cod. Wa, f. 19vb, 20ra: Omnis conceptus qui imprimitur in intellectu possibili ab intellectu agente et phantasmate, vel qui includitur ab eis, est realis. (quoted from T. Barth " De fundamento

logical concept results when the intellect compares one concept with another.[6] The notions of " man " or " animal " are real concepts in the sense that they are abstracted from really existing things. But when one conceives " animal " as a *genus,* or " man " as a *species,* the concept is logical or a second intention, since genus and species express a relationship that one concept bears to another.

From the standpoint of predication, a real concept is predicated directly of things. The concept " desk " is predicated of the desk at which one works. A logical concept (*conceptus rationis*) on the contrary is predicable only of other concepts. " Universal," for instance, cannot be predicated of the actually existing desk, but only of the concept, " desk."

The distinction between real and logical concepts, that is, between first and second intentions, was, of course, the common property of the scholastics. But not all agreed upon the conditions required *a parte rei* before a concept could be called real. The difficult and vexing problem of the reality of the universal is involved. And with Scotus in particular, the objective basis for a real concept is intimately associated with his doctrine of the *natura communis* and the *distinctio formalis a parte rei.*

Both conceptions present difficulties and have been severely criticized both in the Middle Ages and in our own time as unwarranted reifications of ideas.[7] It is not our purpose to discuss

---

univocationis apud Joannem Duns Scotum " in *Antonianum,* XIV (1939), 288. *Oxon.* 1, d. 3, q. 2, n. 8; IX, 19a: Nullus conceptus realis causatur in intellectu viatoris naturaliter, nisi ab his quae sunt naturaliter motiva intellectus nostri, sed illa sunt phantasma vel objectum relucens in phantasmate.

[6] *Prima lectura, loc. cit.:* Sed ille conceptus qui causatur a collatione facta per intellectum alterius conceptus, unius alterius est rationis.

[7] Confer for instance William Ockham, *Ordinatio* I, d. 2, q. 6, EF (revised text by P. Boehner): Contra istam opinionem potest argui duplici via. Primo, quia impossibile est in creaturis aliqua differre formaliter nisi distinguantur realiter, ergo si natura distinguatur aliquo modo ab illa differentia contrahente, oportet quod distinguantur sicut res et res, vel sicut ens rationis et ens rationis vel sicut ens reale et rationis... Secunda

how far either doctrine represents a projection of the logical order into the ontological realm. For the purpose of this study it suffices to call attention to one of the basic reasons which impelled Scotus to postulate both the *natura communis* and the formal distinction.

Scotus was not the originator of the *distinctio formalis*, though he did become one of its greatest champions.[8] There seems to be little doubt that the formal distinction owes its introduction and development to a theological rather than a philosophical controversy, a controversy which centers round the identity-distinction problem of the Blessed Trinity.[9] This controversy, which may roughly be said to have begun with Gilbert de la Porrée, eventuated in a crisis during the 14th century which threatened the validity of Aristotelian formal logic.[10]

Though the present status of historical research forbids us from definitely determining when and in what connection Scotus first definitely adopted the formal distinction, it seem highly probable that it was in connection with the Trinitarian problem. To begin with, it was a question which could not have failed to have its repercussions upon his system. Nor does it seem mere chance that we find the clearest and most detailed treatment of the formal distinction in connection with the question: "Is a

---

via potest argui contra praedictam opinionem, quod non est vera, etiam posito quod esset talis distinctio [i.e. formalis].

[8] B. Jansen, "Beiträge zur geschichtlichen Entwicklung der Distinctio formalis" in *Zeitschrift für katholische Theologie*, LIII (1929), 317ff, 517ff.

[9] Bartholomaeus Roth, *Franz von Mayronis, Sein Leben, seine Werke, seine Lehre vom Formalunterschied in Gott* (Werl, Franziskus-Druckerei, 1936) 283: Die Distinctio formalis fand in die Theologie Eingang als ein Lösungsversuch in der Frage der Bestimmung der Identitätsverhältnisses zwischen persönlichen Proprietäten und Wesenheit, bzw. zwischen Personen und Wesenheit, ferner zwischen persönlichen Proprietäten und Personen und schliesslich zwischen wesentlichen Attributen und Wesenheit in Gott.

[10] Philotheus Boehner, "The Medieval Crisis of Logic and the Author of the Centiloquium attributed to Ockham" in *Franciscan Studies*, XXV (1944), 151-170.

plurality of Persons consistent with a unity of essence?"[11] The very first argument of the *quod non* is the syllogism

Quaecumque uni et eidem simpliciter sunt simpliciter eadem, inter se sunt eadem simpliciter;
Sed personae divinae sunt simpliciter et omnino idem essentiae divinae, quae in se et omnino est simpliciter eadem;
Ergo.[12]

His lengthy solution of the objection is based entirely upon the formal distinction.

But though Scotus probably was attracted to the formal distinction as an answer to a theological problem, it was not long before he had extended it to his metaphysics and psychology. And certainly prompting this further extension, if not actually underlying his Trinitarian arguments, is the further problem of the objectivity of a real concept and the conditions required *a parte rei* for real predication. For there is no doubt that for Scotus the *distinctio formalis a parte rei* seemed the only theory that could safeguard the objectivity of our transcendental notions and make of metaphysics a science of reality.[13]

THE PROBLEM OF PARTIAL KNOWLEDGE

A real concept has to do with really existing things (*res extra animam*). Either it signifies the whole thing or at least something of a thing. The notion of "auto" is real when it signifies and stands for the machine one drives. Likewise the concepts of "wheel," "dashboard," "gear shift," are real, for while they do not stand for or represent the automobile as a whole, they do signify really distinct parts.

---

[11] *Oxon.* 1, d. 2, q. 4, VIII, 503b. Note: questions 4 to 7 of this distinction form a unity. The solutions to the various objections raised in the four questions are given after the body of the 7th question.

[12] *Ibid.* n. 1. Note: the answer to this objection is given in q. 7, n. 47-52; VIII, 630a-634b, after the theory of the formal distinction has been proposed.

[13] *Oxon.* 2, d. 16, n. 17; XIII, 43ab: Isto modo ens continet multas passiones quae non sunt res aliae ab ipso ente... distinguuntur tamen ab invicem formaliter et quidditative, et etiam ab ente, formalitate dico reali et quidditativa; aliter metaphysica concludens tales passiones de ente, et illas considerans, non esset scientia realis.

But Scotus finds difficulty in justifying the validity of partial knowledge when applied to something that is by nature simple, such as God, spiritual substances, etc. When " wisdom " is predicated of God, it is an instance of real predication and " wisdom " is a real concept. Yet wisdom is not a part of God, as a wheel is a part of a wagon. How can my concept be real if it does not grasp the whole of the reality which is God? How can I have partial knowledge of that which has no parts? Or even when such a concept refers to a physical composite like " man ", why is it that our abstractions do not always follow the physical distinctions? Unlike a child clipping paper figures from the magazine page, the mind does not follow the dotted lines of real distinction in carving its concepts from reality. For instance, the very real notions of man as " rational " and " animal " do not correspond to the physical entities of body and soul. Are " rationality " and " animality " mere *entia rationis* projected by the mind into reality?[14] Is man rational, animal and the like, only virtually in the sense that he is capable of producing animal and rational acts? Just as God is only virtually and not formally a stone, in the sense that He is capable of producing the stone in existence?

In fact what is the reality of the whole metaphysical order? What is the objective value of a metaphysical definition, which analyzes the essences of things in terms of generic and differential concepts? What corresponds to those concepts in the thing itself?[15] It is not the thing as a whole, neither is it some physically real part. Is the precise object of the concept a mere *ens rationis*, for after all that is the most a mind can project into a thing?

These problems puzzled Scotus. In his metaphysics he analyzes one of the distinctions suggested by his contemporaries as a possible solution, the *distinctio intentionis*.[16] This " virtual "

---

[14] *Oxon.* 1, d. 8, q. 4, n. 14; IX, 653b: Intellectus actu suo non potest causare nisi relationem rationis.

[15] *Metaphy.* 7, q. 19, n. 4; VII, 465b: Quid istis conceptibus correspondeat in re.

[16] Confer Henry of Ghent's formulation of the *distinctio intentionis.* *Quodl.* V, q. 6, L, 161r: Sed appelatur hic intentio aliquid pertinens realiter

distinction[17] consists in this that the object has a capacity of producing in our mind different concepts of itself. So far as the thing itself is concerned there is no actual distinction nor composition, yet the thing is such that by its very nature (*ipsa nata est facere conceptus*) it tends to give rise to different concepts.[18]

In this the *distinctio intentionis* differs from the purely conceptual distinction in so far as the latter has no intrinsic basis in the thing itself for a difference in concepts.[19]

While such a distinction may account for the difference of concepts, says Scotus, it does not explain how our concepts can be called real, nor how any valid essential or intrinsic predication is possible.[20]

---

ad simplicitatem essentiae alicuius, natum praecise concipi absque aliquo alio a quo non differt re absoluta, quod similiter pertinet ad eandem. *Summa* a. 27, q. 1, n. 25; t. 2, 411:...dico differre intentione quaecumque de se formant diversos conceptus, quorum unus non includit omnino alterum...—Jean Paulus, *Henri de Gand* (Études de Philosophie Médiévale XXV), Paris, J. Vrin, 1938, pp. 220-237.

[17] The term "virtual distinction" is equivocal and is defined differently by different philosophers. Most Scotistic commentators would recognize in the "distinctio intentionis" some form of the virtual distinction. Confer Mastrius de Medula, *Cursus Philosophicus,* IV (*Metaphysic.*) (Venetiis, Apud Nicholaum Pezzana, 1708), p. 318ff, nn. 271-274; M. Grajewski, *The Formal Distinction of Duns Scotus.* (Washington, D. C., Catholic University of America Press, 1944) p. 53.

[18] *Metaphy.* 7, q. 19, n. 4; VII, 465b: Dicunt quidam quod in re sufficit differentia intentionis, quae nullam differentiam nec compositionem actu ponit in re, sed tantum potentialem, sic quod ipsa nata est facere diversos conceptus in intellectu de se. Ita quod ista differentia actu est solum in intellectu concipiente.

[19] *Ibid.;* Non sufficit autem differentia rationis, quae est, quando res non est nata facere, nisi unum conceptum qui tamen potest concipi sub diversis modis concipiendi.

[20] *Ibid.*, n. 5; 465b: Isti conceptus [sc. generis et speciei] videntur fictitii, non reales, nec dicentur *in quid* de specie. *Ibid.*, n. 7; 468a: Et tunc pro opinione de differentia intentionis, est praedicatio ejusdem de se, nec alia erit veritas illarum: *Socrates est homo; Socrates est animal, Socrates est substantia,* etc. Qui melius scit exponere differentiam intentionis, evadendo dictas difficultas, exponat.

Three possibilities are open to us, according to Scotus. Either we grasp the entire reality in each concept, in which case we save the reality of the concept, but we no longer have several concepts but one and the same concept. Or, secondly, we do not grasp reality at all, in which case our notions are pure mental fictions or constructs. Or, finally, the mind does not grasp the entire reality, but only something of reality (*aliquid rei*). But this last hypothesis implies that some sort of actual difference exists in the thing prior to any operation of the mind. At least to the extent that *a parte rei* that which corresponds precisely or formally to one concept is not completely identical with that which is represented in the other.[21]

> In such notions as these, does the intellect, I ask, have as the object of its knowledge something existing in the thing? If not, we have a mere mental fiction. If it is the same thing, the object of both concepts is identical unless you admit that one and the same extra-mental thing formally generates two objects in the intellect. And in this case, it does not seem that the thing or anything of the thing is the object of my knowledge, but rather something produced by the thing. But if the intellect knows something different in each concept, then the thesis is granted, since a difference exists prior to the concepts.[22]

As an alternate solution he suggests what is to all appearances the formal distinction *a parte rei*. This opinion does not deny

[21] *Ibid.* n. 5; 465b: Sed quod nec differentia ista intentionis sufficit, arguitur sic: quia concipiendo genus, aut concipitur aliquid rei in specie, aut nihil, similiter de differentia; si nihil, isti conceptus videntur fictitii, non reales, nec dicentur in *quid* de specie; si aliquid, aut aliquid idem, et tunc erit idem conceptus; aut aliquid aliud, et tunc erit in re aliqua differentia prior differentia conceptuum.

[22] *Ibid.* 466a: Quaero igitur, an istis notitiis cognoscat intellectus objective aliquid in re? Si nihil, fictio est; si idem, ergo objectum idem est, nisi dicas quod una res extra facit formaliter duo objecta in intellectu, et tunc non videtur quod res vel aliquid rei sit objectum, sed aliquid factum a re; si aliud, habetur propositum, quia differentia ante conceptus.

the *distinctio intentionis* outright, but rather places the necessary conditions *a parte rei* for its validity.[23]

THE FORMAL DISTINCTION

This approach to the formal distinction indicates that one of its greatest attractions for Scotus lay in the fact that it provided an objective basis for our real concepts. And the concepts he had in mind were not merely those of genus and specific difference, but what is of particular interest for us, the concepts of being and its attributes as well as those relative attributes in God and creatures which play such an important part in metaphysics as a theologic. "It seems necessary to posit this distinction in other things, for example, in being and its attributes, in the relation and its foundation in God and creatures." [24]

This formal distinction is something less than a real physical distinction (*realis simpliciter*) which exists between two or more physical entities (*inter rem et rem*).[25] At the same time it is not a mere distinction created by the mind (*distinctio rationis*).[26]

---

[23] *Ibid.* n. 10; 470a; Ista opinio non negat differentiam intentionis, sed ponit sibi necessaria correspondere aliquam in re.—Note that Scotus also calls the formal distinction a virtual intrinsic distinction. Confer note 30. See also where Scotus identifies a formality with the *intentio* in so far as the latter is conceived as an aspect of reality. *Rep. Par.* 2, d. 1, q. 6, n. 20; XXIII, 556b: Intelligit idem ipse Avicenna per aliam intentionem quod ego dico per aliam formalitatem.

[24] *Metaphy.* 7, q. 19, n. 8; VII, 468b: Istam differentiam videtur necessarium ponere in aliis, puta in ente et ejus passionibus, relatione et fundamento in Deo et creatura.

[25] *Oxon.* 1, d. 2, q. 7, n. 43; VIII, 603a: Illud quod habet talem [i.e. formalem] distinctionem in se non habet rem et rem, sed est una res habens virtualiter sive eminenter quasi duas realitates, quia utrique realitati ut est in illa re, competit illud proprium quod inest tali realitati, ac si ipsa esset res distincta, ita enim haec realitas distinguit, et illa non distinguit, sicut si ista esset una res et illa alia.

[26] *Ibid.* n. 42; 598a: Igitur quaecumque intrinseca sunt diversa objecta formalia intuibilia, secundum propriam existentiam actualem terminant intuitionem ut objecta et ita habent aliquam distinctionem ante actum intelligendi.

It is real in the sense that the mind discovers it but does not project it into reality (*realis secundum quid*).[27] It exists between *rationes reales* or *formalitates*, and not between *res et res*.[28]

The properties of these *formalitates* might be summed up as follows:

1. The formality is not a distinct physical thing, but a positive something that is somehow less than a thing.[29] It is the *ratio objectiva* of a distinct formal concept.[30]

---

[27] *Ibid.* n. 41; 597a: Et intelligo sic realiter, quod nullo modo per actum intellectus considerantis, imo quod talis entitas esset ibi si nullus intellectus consideraret, et sic esse ibi si nullus intellectus consideraret, dico esse ante omnem actum intellectus. *Report. Par.* 1, d. 45, q. 2, n. 9ff; XXII, 502ff.

[28] *Oxon. l.c.* n. 45; 604a: Ista differentia manifestatur per exempla, primo si ponatur albedo species simplex, non habens in se duas naturas, est tamen in albedine aliquid realiter unde habet rationem coloris, et aliquid unde habet rationem differentiae et haec realitas formaliter non est illa realitas, nec e converso formaliter; imo una est extra realitatem alterius, formaliter loquendo, sicut si essent duae res, licet modo per identitatem istae duae realitates sint una res.

[29] *Oxon.* 2, d. 3, q. 6, n. 15; XII, 144a: Quodlibet commune, et tamen determinabile adhuc potest distingui, *quantumcumque sit una res*, in plures realitates, formaliter distinctas, quarum haec formaliter non est illa. (Italics mine) The classical expression of the Scotists for a formality is that it is not *res sed rei*, or it is an *aliquitas rei*. Scotus himself refers to it sometimes as *aliquid rei*. Confer *Metaphy.* 7, q. 19, n. 5; VII, 466a.

[30] In fact, from this standpoint it might in an equivocal sense be called a *distinctio rationis*. Confer *Oxon.* 1, d. 2, q. 7, n. 43; IX, 603a: Potest autem vocari differentia rationis, sicut dicit Doctor quidam, non quod *ratio* accipiatur pro differentia formata ab intellectu, sed ut ratio accipitur pro quidditate rei secundum quod quidditas est objectum intellectus; vel alio modo potest vocari differentia virtualis, quia illud quod habet talem distinctionem, in se non habet rem et rem, sed est una res habens virtualiter sive eminenter quasi duas realitates, quia utrique realitati ut est in illa re, competit illud proprium quod inest tali realitati, ac si ipsa esset res distincta, ita enim haec realitas distinguit, et illa non distinguit, sicut si ista esset una res et illa alia. Some Scotists define formalities in terms of a *ratio objectiva*. See for instance B. Sanning, *Scholae Philosophicae Scotistarum* seu *Cursus Philosophici ad mentem Doctoris Subtilis Joannis Duns Scoti* (Neo-Pragae, typis Hampelii, 1685) I, 126: Formalitas est ratio objectiva identificata rei, conceptibilis in re aliqua conceptu adaequato, et perfecto, distincto a conceptu quo concipitur alia formalitas ejusdem rei.

2. Each formality has its own proper quiddity or entity. Though the simplicity of the formality may forbid a strict metaphysical definition, if formalities could be defined, their definitions would not be simply synonymous but would differ essentially.[31]

3. Since these formalities do not have a distinct existence, but rather exist by the existence of the thing, they are inseparable even by the power of God. Thus, while God could annihilate the rational soul, or could create an irrational soul, He could not separate " sensitivity " or " rationality " from the human soul. The formality consequently is not the product of a distinct physical causality over and above the causality which brings the thing as a whole into existence.[32] For these perfections or formalities are *unitive contentae,* in the sense that they constitute one indivisible *res.*[33]

4. Just as we may speak of a real distinction between the whole and its parts, so we may speak of a formal distinction between the thing as a whole and the single formalities. Such

---

[31] *Oxon.* 1, d. 8, q. 4, n. 18; IX, 665a: Quod autem non includat formaliter ut in communi, hoc declaro, quia includere formaliter est includere aliquid in ratione sua essentiali, ita quod si definitio includentis assignaretur, inclusum esset definitio vel pars definitionis. Sicut autem definitio bonitatis in communi non habet sapientiam in se, ita nec infinita infinitam. Est igitur aliqua non identitas formalis sapientiae et bonitatis, inquantum earum essent distinctae definitiones, si essent definibiles; definitio autem non tantum indicat rationem causatam ab intellectu, sed quidditatem rei, ergo non est identitas formalis ex parte rei.

[32] *Oxon.* 2, d. 3, q. 4, n. 6; XII, 95a: Quamlibet enim entitatem consequitur propria unitas, non habens aliam causam propriam sui quam causam entitatis.

[33] *Rep. Par.* 4, d. 46, q. 3, n. 4; XX, 448a: Unitive autem non continentur quae sine omni distinctione continentur, quia unio non est absque omni distinctione. Nec unitive continentur quae simpliciter realiter distincta continentur, quia illa multipliciter, seu dispersim continentur; ergo hoc vocabulum *unitive* includit aliqualem distinctionem contentorum, quae sufficit ad unionem; non tamen talem unionem quae repugnet omni compositioni et aggregationi distinctorum, hoc non potest esse nisi ponatur non identitas formalis cum identitate reali. *Oxon.* 1, d. 8, q. 4, n. 18; IX, 665a.

a distinction is called inadequate. Such an inadequate formal distinction exists, for instance, between " animality " and " sensitivity ", for the former includes the latter.

THE MODAL DISTINCTION

Besides the strict formal distinction existing between two or more formalities, Scotus introduces a similar distinction between a formalty and its intrinsic mode.[34] Such, for instance, is the distinction between intelligence and the modality of finiteness in man, or between any divine perfection, like wisdom and its mode of infinity. Scotists dispute whether this so-called *distinctio formalis modalis* is really a distinction *a parte rei* or merely a virtual or mental distinction with an extrinsic foundation in things.[35] While reasons for both interpretations may be found, the author is inclined to believe that Scotus regarded it as a distinction *a parte rei* but less than the strict formal distinction. Even within the limits of the strict formal distinction Scotus admits of various gradations.[36] The line between the least of the formal distinctions and this " modal " distinction is easily crossed, if such a line exists at all.

The modal distinction seems to be *a parte rei,* for Scotus refers to it as a *distinctio in re* [37] and regards it as the foundation for

---

[34] *Oxon.* 1, d. 8, q. 3, n. 27; IX, 626b-627b.

[35] Extrinsic is here contrasted with an intrinsic foundation. The former gives rise to a mental distinction, the latter to the formal distinction *a parte rei.* Hence the formal distinction is sometimes called an *intrinsic* virtual distinction. Extrinsic foundation simply means that the basis of the distinction is extrinsic to the thing distinguished. For instance, on the basis of a difference in acts, man's intellectual faculty is distinguished as intellect and as reason. As effects of the mind, these acts are really distinct accidental modes and hence extrinsic to the mind itself.

[36] *Metaphy.* 7, q. 19, n. 8; VII, 468b: Est enim maxima naturarum et suppositorum. Media naturarum in uno supposito. Minima diversarum perfectionum, sive rationum perfectionalium unitive contentarum in una natura.

[37] *Oxon.* 1, d. 8, q. 3, n. 27; IX, 627a:...distinctio in re sicut realitas et sui modi intrinseci...

those real, though imperfect, concepts which are predicable univocally of God and creatures. The assertion that such notions as being, wisdom, free will, are real concepts predicable of God gives rise to the same problem as did the genus and specific difference. "How can the concept which is common to God and creatures be considered real unless it be abstracted from some reality of the same kind?" [38] The perfection and its intrinsic mode, Scotus answers, are not so identical that we cannot conceive the perfection without the mode. In other words, the perfection and its mode are not perfectly identical—not indeed in the sense that there is any possibility of separating the two (real distinction), nor even in the sense that both the perfection and the mode are each capable of terminating a distinct and proper concept (strict formal distinction), but only to the extent that the objective reality signified by a concept which includes the mode is not precisely the same as that which is signified by a concept which does not include the mode. Such a distinction in the thing suffices to safeguard the reality of those imperfect and common concepts predicable of God and creatures.

This is not yet a strict formal distinction because an intrinsic mode is not a formality in its own right. As will be discussed later, such a mode is essentially a qualification. It includes both in thought and in definition the notion of the subject of which it is the mode, even though the subject enters the definition *ἐκ προσθέσεως*, as Aristotle put it.[39] The mode consequently is incapable of terminating a distinct and proper concept. With the perfection which it modifies the case is slightly different. It can be conceived without including the modality at all. But such a concept is imperfect. It does not give the full perfection of the formality in question. For instance, when we conceive God as a being, or as wise, we are using notions that are common to creatures. Yet these perfections as they actually exist in God are formally infinite. So much so that if we were gifted with

---

[38] *Ibid.;* 626b: Sed hic est unum dubium. Quomodo potest conceptus communis Deo et creaturae realis accipi, nisi ab aliqua realitate ejusdem generis?

[39] Confer chapter 4, pp. 88-89.

the intuitive knowledge of the blessed in heaven, we should not perceive the perfection of wisdom, for instance, and the modality of infinity as two distinct formal objects but only as one.[40] When

[40] *Oxon.* 1, d. 8, q. 3, n. 27-29; IX, 626b-628b: Quando intelligitur aliqua realitas cum modo intrinseco suo, ille conceptus non est ita simpliciter simplex, quin possit concipi illa realitas absque modo illo, sed tunc est conceptus imperfectus illius rei; potest etiam concipi sub illo modo, et tunc est conceptus perfectus illius rei.... Requiritur igitur distinctio inter illud a quo accipitur conceptus proprius non ut distinctio realitatis et realitatis, sed ut distinctio realitatis, et modi proprii et intrinseci ejusdem; quae distinctio sufficit ad habendum perfectum conceptum vel imperfectum de eodem, quorum imperfectus sit communis et perfectus sit proprius; sed conceptus generis et differentiae requirunt distinctiones realitatum, non tantum ejusdem realitatis perfecte et imperfecte conceptae.

Istud potest sic declarari: Si ponamus aliquem intellectum perfecte moveri a colore ad intelligendum realitatem coloris et realitatem differentiae, quantumcumque habeat perfectum conceptum adaequatum primae realitati, non tamen habet in hoc conceptum realitatis a quo accipitur differentia, nec e converso, sed habet ibi duo objecta formalia, quae nata sunt terminare conceptus proprios distinctos. Si autem tantum esset distinctio in re sicut realitatis et sui modi intrinseci, non posset intellectus habere proprium conceptum illius realitatis, et non habere conceptum illius modi intrinseci rei, saltem ut modi sub quo conciperetur, licet ille modus non conciperetur, sicut de singularitate concepta, et modo sub quo dicitur concipi, ut patet alibi, sed in illo perfecto conceptu haberet unum objectum adaequatum illi, scilicet rem sub modo.

Et si dicas, saltem conceptus communis est indeterminatus et potentialis ad specialem conceptum, ut realitas ad realitatem, vel saltem non erit infinitus, quia nullum infinitum est potentiale ad aliquid. Concedo quod conceptus ille communis Deo et creaturae est finitus, hoc est, non est de se infinitus positive, ita quod de se includat infinitatem, quia si esset infinitus non esset communis de se finito et infinito; nec est de se finitus positive, ita quod de se includat finitatem, quia tunc non competeret infinito, sed est de se indifferens ad finitum et infinitum, et ideo est finitus negative, id est, non ponens infinitatem, et tali finitate est determinabilis per aliquem conceptum...

Sed si arguas, ergo realitas a qua accipitur, est finita, non sequitur; non enim accipitur ab aliqua realitate, ut conceptus adaequatus realitati, sive ut perfectus conceptus illi realitati adaequatus, sed ut diminutus vel imperfectus. In tantum etiam quod si illa realitas a qua accipitur, videtur perfecte et intuitive, intuens ibi non haberet distincta objecta formalia, scilicet realitatem et modum, sed idem objectum formale; tamen intelligens intellectione abstractiva, propter intellectionis illius imperfectionem potest unum habere pro objecto formali, licet non habeat alterum.

the perfection is conceived together with its mode, the concept is said to be proper; conceived without its mode, the concept is imperfect and common. Where the strict formal distinction obtains, two distinct formal objects are to be had. Each is capable of terminating a proper concept.[41] Thus God's will is not proper to His intellect, nor vice versa in the sense that the modality of infinity is proper to His intellect or to His will. Even the intuitive knowledge of the blessed will not erase the formal difference of the two.[42]

THE "NATURA COMMUNIS"

Scotus posits some kind of formal, or at least modal, distinction *a parte rei* between being and its transcendentals,[43] between the various metaphysical grades of being,[44] between the metaphysical essence and its properties,[45] and in the realm of psychology,

---

[41] *Ibid.,* n. 27; 627a.

[42] *Metaphy.* 7, q. 19, n. 7; VII, 467b-468a: Non videtur etiam, quod aliqua intellectionum istarum aliquid rei cognoscatur. Quod ostenditur sic: quia sic intellectioni abstractivae potest succedere intuitiva ejusdem primi objecti praecise, quando abstractiva alicujus rei est imperfecta, alia perfecta; sed visiones duae possunt esse generis et differentiae, quia visio non est nisi objecti primi realiter existentis et praesentis. Ex hoc sequitur quod intellectus divinus non cognoscit, ut distincta objecta ista, quorum ponitur differentia intentionis, nec aliquis intellectus, nisi abstractive intelligens, et ita imperfecte.

[43] *Oxon.* 2, d. 16, q. un. n. 17; XIII, 43a: Alia sunt contenta in aliquo unitive quasi posteriora, quia quasi passiones continentis, nec sunt res aliae ab ipso continente. Isto modo ens continet multas passiones, quae non sunt res aliae ab ipso ente, ut probat Aristoteles in *princip.* 4 *Metaphy.* distinguuntur tamen ab invicem formaliter et quidditative, et etiam ab ente, formalitate dico reali et quidditativa. (Note that this passage is not in the *Ordinatio* proper.) Confer also *Oxon.* 3, d. 8, q. un, n. 17; XIV, 377a: Nec ista contradicunt, licet enim entitati absolute sit idem realiter veritas et bonitas, et huic entitati haec veritas et haec bonitas, non tamen formaliter et quidditative, quia veritas et bonitas sicut quasi passiones entis; *Collatio* 36, n. 4; V, 297b-298a.

[44] *Metaphy.* 7, q. 19; VII, 462b: Utrum conceptus generis sit alius a conceptu differentiae?

[45] By the properties, we mean those attributes or *passiones* which are predicated of their subject (the essence) in the second mode of *per se*

between the faculties of the soul and the soul itself,[46] besides the theological applications [47]—which in Scotus' eyes—are its primary justification.

Over and above these instances, Scotus makes a very important application of the formal distinction in his theory of the *natura communis.* This doctrine will be touched upon in a latter chapter.[48] It should be remarked here, however, that it is this same tendency to find an objective basis for our concepts that prompted Scotus to postulate as formally distinct from the individuating difference or *haecceitas,* a common nature (*natura communis*) which serves as the *fundamentum in re* for the universal concept. In this sense there is a definite connection be-

---

predication, for instance, *one, good* of being, *risible* of man. Such attributes are contained in the thing *quasi posteriora* in contradistinction to the grades distinguishable in the metaphysical essence, which are *quasi superiora.—Oxon.* 2. d, 16, q. un. n. 17; XIII, 43a-44a: Ideo dico aliter sic, secundum Dionysium 5 c. *de Divin. nom.* continentia unitiva non est eorum quae omnino sunt idem, quia illa non uniuntur, nec est eorum quae manent distincta, ista distinctione, qua fuerunt distincta ante unionem; sed quae sunt unum realiter, manent tamen distincta formaliter, sive quae sunt idem identitate reali, distincta tamen formaliter; hujusmodi autem contenta sunt in duplici differentia, quia quaedam sunt de natura continentis, ut quaecumque sunt superiora ad continens, v.g. ab eadem re accipitur ratio albedinis, coloris, qualitatis sensibilis, et qualitatis, et haec sunt superiora ad hanc albedinem, et ideo omnia sunt de essentia ejus. Alia sunt contenta in aliquo unitive quasi posteriora, quia quasi passiones continentis, nec sunt res aliae ab ipso continente. Isto modo ens continet multas passiones, quae non sunt res aliae ab ipso ente...distinguuntur tamen ab invicem formaliter et quidditative, et etiam ab ente...Sic ergo possumus accipere de intellectu et voluntate, quae non sunt partes essentiales animae, sed sunt unitive contenta in anima quasi passiones ejus, propter quas anima est operativa, non quod sint essentia ejus formaliter, sed sunt formaliter distinctae, idem tamen identice et unitive...

[46] *Ibid.*

[47] Besides the most important application of the distinction to the Persons and Nature in the Blessed Trinity, Scotus also applies the distinction to the divine attributes. Confer *Oxon.* 1, d. 8, q. 4, IX, 636ff.; B. Roth, *Franz von Mayronis,* passim.

[48] Confer chapter 5, pp. 107-111.

tween the doctrine of the formal distinction and the *natura communis* as J. Kraus has indicated.[49]

Scotus could not accept without reservation the Aristotelian dictum, *Intellectus est universalium, sensus singularium.*[50] As a Christian metaphysician, looking forward someday to the beatific vision, he had to qualify it substantially. Granting that singularity *qua* singularity is not intelligible to our intellect in its present state, it is intelligible as such, and will be known one day by our intellect.[51] Even in this life we know singular natures and know that they are singular. We do not perceive the precise formal reason why they are singulars.

For this reason all our clear and distinct concepts, in terms of which we formulate our definitions of the nature and essence of things, are only partial conceptions. They give us not the concrete individual in its full intelligibility, but only the common elements—those *rationes* which it has in common, or could have in common, with other individuals. And to safeguard the objectivity of these notions, upon which all scientific knowledge is based, Scotus postulated the *natura communis*. Formally distinct from its principle of individuation, endowed with its own proper unity, which is somehow less than numerical unity, this *natura communis*, as actualized in concrete individual things, forms the immediate and proper object of the *conceptus realis*.[52]

---

[49] Johannes Kraus, *Die Lehre des Johannes Duns Skotus von der Natura Communis* (Freiburg, Schweiz, Studia Friburgensia, 1927) 142: Einen Zusammenhang möchten wir aber insofern annehmen, als wir glauben, dass hier wie dort die gleiche Systematik, der gleiche Leitgedanke, sich auswirkt. —Kraus, however, has minimized the importance of the formal distinction in this connection.

[50] Confer Aristotle, *De Anima*, II, 5 (417b 22): αἴτιον δ'ὅτι τῶν καθ' ἕκαστον ἡ κατ' ἐνέργειαν αἰσθήσις, ἡ δ' ἐπιστήμη τῶν καθόλου; *Analytica Post.* I, 24 (passim); Scotus, *Report. Par.* 3, d. 14, q. 3, n. 9; XXIII, 358a; *Oxon.* 4, d. 45, q. 3, n. 17; XX, 348b.

[51] *Oxon.* 2, d. 3, q. 9, n. 9; XII, 276b: Et cum dicis, quod intellectus noster non intelligit singulare, primo dico, quod hoc est imperfectionis pro statu isto. *Metaphy.* 7, q. 15: Utrum singulare sit per se intelligibile a nobis? (VII, 434ff).

[52] Confer J. Kraus, *op. cit.*, pp. 136ff; P. Minges, "Der angebliche exzessive Realismus des Duns Scotus" in *Beiträge zur Geschichte der Philosophie*

Applying these distinctions to the transcendentals, it can be said that, according to Scotus, corresponding to each distinct transcendental notion is some real perfection or mode. While such realities or perfections cannot be separated physically from one another but are separable only in thought, it must be admitted that, prior to any operation of the mind, that phase of reality which a mind could grasp by real though partial knowledge is not formally identical *a parte rei* with the remaining reality which it fails to grasp. Such a distinction can be called real only *secundum quid.* As such, it does not imply any imperfection or composition, yet it suffices to guarantee the reality of the metaphysical order.

The formal and modal distinction is not easy to grasp. It raises many difficulties which neither Scotus nor his followers fully solved. Perhaps the most intelligible and still fairly accurate notion of a " formality " is to consider it as the objective basis of a concept which, though real, does not represent the whole intelligible content of the physical entity, but a part only.

While it is possible to form distinct real concepts only where some sort of formal non-identity exists *a parte rei,* Scotus nowhere implies that the mind cannot grasp many such " formalities " in one and the same simple concept. Certainly this is possible for God and also for a creature gifted with intuitive vision. But even with such knowledge, the human intellect would recognize, for instance, that God's intelligence is not formally identical with His free will, or that the intelligibility of the divine essence is not formally identical wth its appetibility, and so on. In a word, the mind would recognize that a distinct real concept of one such perfection without the other is possible.

The formal distinction, in short, is not due solely to the imperfection of our knowledge, though the possibility of separating in thought such non-identical perfections *a parte rei* instead of grasping them in *uno actu,* as God does, is due to the imperfection of our created minds.

---

des Mittelalters, VII (Münster 1908), pp. 81ff; Scotus, *Oxon.* 2, d. 3, q. 1, etc.

CHAPTER III

UNIVOCATION AND TRANSCENDENTALITY

THE study of the formal and modal distinctions in the previous chapter reveals to what extent the transcendentals of metaphysics may be called real notions. An analysis of Scotus' doctrine of univocation is very helpful in understanding the transcendentals in so far as they are concepts. Univocation is not just an incidental factor in his conception of transcendentality. In fact it might be considered the very foundation of his theory. It solves the paradox of being; it underlies the law of the disjunctive transcendentals; it pervades the whole realm of the pure perfections. Even more important in the eyes of Scotus, it is a *sine qua non* assumption in any Aristotelian metaphysics aspiring to become a *theologia de Deo*.[1]

Unfortunately, the subject of univocity and analogy became a storm center of controversy between the Thomistic and Scotistic schools of thought. For the dispute, far from clearing the atmosphere, has perhaps been more successful in obscuring the real position of Duns Scotus and, we suspect, of St. Thomas himself.[2] For those who regard Scotus' theory of univocity as

[1] *Oxon.* 1, d. 3, q. 1, n. 2; IX, 8b: Metaphysica est Theologia de Deo.—Confer Aristotle, *Metaphysica* VI, 1 (1026a 18ff).

[2] A careful analysis of the positions of Duns Scotus and of St. Thomas in the light of their own statements and without the benefit of well-meaning commentators indicates a far more fundamental agreement between the two men than the superficial treatment characteristic of the average neo-scholastic textbooks, or even reference works, would lead one to believe. Not that the two doctrines are simply identical. Each of the two had his own approach and his own problems. It is the personal opinion of the author that the doctrine of St. Thomas and that of Duns Scotus are fundamentally compatible. But since this is itself a controversial question and involves a criticism of certain phases of neo-Thomism, we have refrained from any positive attempt to correlate the two.

a direct challenge to the position of St. Thomas, it might be interesting to note that the Franciscan master apparently did not even have his illustrious predecessor in mind. What he really challenged was the Augustinian illumination theory and, more specifically, the modified form it had assumed in the system of Henry of Ghent.

The significance of this starting point has been almost universally overlooked,[3] yet it is really the key to comprehending Scotus' position. For like most of his contemporaries he had definitely broken with this theory of knowledge.[4] Scotus, however, was one of the first, if not the first, clearly to perceive the full implications of that break. The real roots of the medieval theory of analogy are to be found in Augustinian illuminationism, or even more remotely, perhaps, in Plato himself.[5] In rejecting illuminationism, Scotus was forced to find a new basis for the doctrine of analogy. He found it in his theory of univocation.

---

[3] A noteworthy exception in this regard is P. Boehner. Confer his *History of Franciscan Philosophy* (manuscript) p. 68. Another is J. Paulus, *Henri de Gand* (Paris, J. Vrin, 1938), p. 63ff.

[4] Confer Scotus' criticism of the Augustinian theory of illumination in general and that of Henry of Ghent in particular, *Oxon.* 1, d. 3, q. 4; IX, 162ff.

[5] Contemporary neo-scholastics are beginning to rediscover the difficulty of harmonizing the Aristotelian theory of ideogenesis with the hypothesis of a purely analogical knowledge of God. In this connection it is interesting to quote the following observation from a recent article.

"Darüber dürfen die Schwierigkeiten nicht übersehen werden. Es sind vornehmlich zwei. Einmal ergibt sich hier eine axiologische Parallele zu dem zweifellos schwierigen ontologischen Problem der analogen Prädikation des Seins von der Kreatur und von Gott, dem bedingten und dem unbedingten Sein. Platon hat nicht nur einmal zu erkennen gegeben, dass er nicht sagen kann, was das an sich Gute wäre, wenn es nicht ein Gut für etwas sein soll. Und dann besteht noch die andere ebenso tiefgreifende Frage: Wie kommen wir zu der Erkenntniss und der Evidenz der Idealgestalt eines Seinenden, wenn wir nicht wie Platon an der Anamnesis und der daran sich entfaltenden Dialektik festhålten, sondern unseren Aussagen über das Seiende eine synthetische Erkenntnis zu Grunde legen?" Johannes Hirschberger, "Omne ens est bonum" in *Philosophisches Jahrbuch*, LIII (1940), 305.

THE SO-CALLED "ANALOGICAL CONCEPT"

To realize the full import of the problem Scotus had to face, a word regarding his conception of analogy and univocation is required. We do not intend to go very extensively into his theory of analogy,[6] but in view of the descriptions found in the average scholastic reference work or textbook, it is necessary to say a word about the unity of the analogical concept.[7]

A concept is a formal sign of a thing—formal in the sense that its whole being is to signify. Its meaning, its content, so to speak, is its signification. It is psychologically impossible therefore for one and the same concept to have two different meanings. Where a given predicate has different significations depending upon the subject of which it is predicated, the predicate obviously is one and the same only in name. There are as many different concepts as there are different significations. Whether the meaning is similar or not does not alter this fact. Speaking freely, if somewhat inexactly, it is said that the same notion or formal thought-content (*ratio*) is predicated now one

---

[6] Scotus in fact does not have any lengthy discussion of analogy in his more important works since it plays only a minor role in his system.

[7] It is rather difficult to understand just what is meant by the phrase "unity of the analogical concept." Even the neo-scholastics who speak of such a thing do not seem to be agreed upon it. The impression is sometimes created that just as one and the same term can have several significations, so one and the same concept can have different meanings. See, for instance, Pedro Descoqs, S.J. *Praelectiones theologiae naturalis* (Paris, Gabriel Beauchesne et Ses Fils, 1935) II, 752: "*Analogice* tribuitur pluribus *ille terminus seu conceptus, qui dicitur de pluribus secundum rationem eandem et diversam,* h. e. ratio quae nomine communi importatur et conceptu uno repraesentatur, tribuitur vel essentiis vel subjectis quae sunt in se vere diversa, quae tamen inter se manifestant aliquam similitudinem et habitudinem sive in esse, sive in operatione atque ideo hujusmodi sunt ut conceptu uno repraesentari valeant." Confer also W. E. Byles, "The Analogy of Being" in *The New Scholasticism,* XVI (1942), p. 346f, V. E. Smith, "On the 'Being' of Metaphysics," *Ibid.* XX (1946), 81ff. Gredt has one of the more intelligible explanations. The "ratio significata" is simply different in each case, but because of some relation between the "rationes significatae" they are said to be one only in a qualified sense (*secundum quid*). *Elementa Philosophiae Aristotelico-Thomisticae* (Freiburg im Breisgau, Herder, 1937) I, 131.

way, now another. In actual predication, however, the predicate has but one meaning. Either the mere formal thought-content is predicated with no intent to signify the mode of existence or both the formal *ratio* and the mode of existence are signified. But in the latter instance we no longer have the same concept, nor, for that matter, the same formal thought-content.[8] In so-called " analogical predication "—to use an un-Aristotelian expression—the predicate is not a common concept at all but a common name to which correspond two distinct, though related, concepts.

Whenever a common name is predicated essentially of several things, as Aristotle tells us in the opening chapter of the *De Categoriis,* they are said to be named univocally or equivocally, depending upon whether the notion (λόγος τῆς οὐσίας) expressed by the name is common or different.[9] The λόγοι τῆς οὐσίας may be related to one another, for instance, by attribution or proportion or similarity, as Boethius indicates.[10] But they still remain equivocal from the viewpoint of a logician. Only if it is possible to prescind completely from the differences between two things and to grasp only the common element intrinsic to both (λόγος τῆς οὐσίας), does the term become univocal.

The logician consequently knows nothing of analogy or analogical concepts as a mean between pure equivocation and univocation.[11] Analogy exists only for the metaphysician, namely, where the two objects of comparison (whether they be things

---

[8] St. Thomas, incidentally, is very clear on this point. *Summa theologiae,* I, q. 13, a. 5, c.: Neque enim in his quae analogice dicuntur, est una ratio, sicut est in univocis; nec totaliter diversa, sicut in aequivocis; sed nomen quod sic multipliciter dicitur, significat diversas proportiones ad aliquid unum.

[9] *De Categoriis,* 1 (1a 1-12).

[10] Boethius, *In Categorias Aristotelis Libri Quatuor,* lib. 1 (PL 64, 166 B).

[11] *Oxon.* 1, d. 8, q. 3, n. 14; IX, 592b: Univocum et aequivocum sunt immediata apud Logicum. *Super lib. Elenchorum,* q. 15, n. 7; II, 22b: Inter idem et diversum non cadat medium, ideo Logicus medium non ponit inter aequivocum, et univocum.

or concepts) are not simply diverse but related in such a way that one may be called by the name of the other.[12]

Modern logicians tend simply to reduce all forms of analogy to the *aequivocatio a consilio*.[13] In his commentary on the Categories of Aristotle, Boethius, influenced probably by the *De Sophisticis Elenchis*,[14] distinguishes between *aequivoca a casu* and *aequivoca a consilio*. In the case of the first, it is a mere coincidence that two different things have been given the same name, for instance, " logistics " as applied to a branch of military art and to a branch of logic. In the *aequivoca a consilio*, however, the choice of the name is deliberate or intentional and is based upon some similarity or proportion or other relation that justifies the attribution of a common name to both.[15]

---

[12] *Metaphy.* 4, q. 1, n. 12; VII, 153a: Aequivoce dicitur aliquid de multis, quando illa, de quibus dicitur non habent attributionem ad invicem, sed quando attribuuntur, tunc analogice.... et simpliciter aequivoce secundum Logicum... analogice secundum Metaphysicum realem. *Super Praed.* q. 4, n. 7; I, 447b.

[13] Confer for instance, J. Gredt, *Elementa Phil. Arist.-Thomist.*, I, 131: " Si ratio significata, simpliciter diversa, aliquomodo tamen est una seu una secundum quid, etiam terminus aequivocus est *secundum quid* tantum seu aequivocus *a consilio*." W. Esdaile Byles, " The Analogy of Being " in *The New Scholasticism*, XVI (1942), p. 336, note 20: " What we call ' analogy ' Boethius calls ' consiliar equivocity,' pure equivocation being casual equivocity"; H. C. Plassmann, *Die Schule des hl. Thomas v. Aquino* (Soest, Verlag der Nasse'schen Buchhandlung, 1858) V, 307: " Das *aequivocum* theilt sich in ein *aequivocum a casu* und *aequivocum a consilio*. Ersteres, das zufällig äquivoke, ist das schlechthin und mit unserer definition gemeinte Aequivoke. Letzteres, das absichtlich äquivoke, ist nicht schlechthin ein äquivokes Ding, sondern eben unser *analogum*."

[14] *De Sophisticis Elenchis*, c. 4 (166a 15ff).

[15] Boethius, *In Categorias* I (PL 64, 166 B): Aequivocorum alia sunt casu alia consilio. Casu, ut Alexander Priami filius et Alexander Magnus, Casus enim id egit, ut idem utrique nomen poneretur. Consilio vero, ea quaecunque hominum voluntate sunt posita. Horum autem alia sunt secundum similitudinem, ut homo pictus et homo verus quo nunc utitur Aristoteles exemplo: alia secundum proportionem, ut principium est in numero unitas, in lineis punctus. Et haec aequivocatio secundum proportionem esse dicitur. Alia vero sunt quae ab uno descendunt ut medicinale ferramentum; medicinale pigmentum, ab una enim medicina aequi-

Aristotle, St. Thomas tells us, conceived equivocation in a broad sense as including analogy.[16] From the viewpoint of formal logic he had no other alternative, for even terms that are equivocal *a consilio* invalidate a syllogistic argument and are included among the three classes of equivocal or ambiguous words to be avoided by the logician.[17]

DEFINITION OF UNIVOCAL CONCEPT

With this is mind we can understand why Scotus defined univocity as he did. Lest there be a mere quibble about terms, he says, by a univocal concept I mean one that possesses such a unity or singleness of meaning that it will serve as the middle term of a valid syllogism, or if one were to affirm and deny it simultaneously of one and the same thing, the result would be a contradiction.[18]

Consider the following syllogism:

> Whatever is divine is God;
> But the Mosaic law is divine;
> Therefore the Mosaic law is God.

The syllogism is obviously a fallacy of equivocation, for "divine" is used in an analogous sense when applied to the Mosaic law. In the major premise it expresses an intrinsic

---

vocatio ista descendit. Alia quae ad unum referuntur, ut si quid dicat salutaris vectatio est, salutaris esca est, haec scilicet idcirco sunt aequivoca, quod ad salutis unum vocabulum referuntur.

[16] *Summa theol.* I, q. 13, a. 10, ad 4: Philosophus largo modo accipit aequivoca, secundum quod includunt in se analoga. Quia et ens, quod analogice dicitur, aliquando dicitur aequivoce praedicari de diversis praedicamentis.

[17] *De Sophisticis Elenchis*, c. 4 (166a 15ff).

[18] *Oxon.* 1, d. 3, q. 2, n. 5; IX, 18a: Et ne fiat contentio de nomine univocationis, conceptum univocum dico, qui ita est unus, quod ejus unitas sufficit ad contradictionem, affirmando et negando ipsum de eodem. Sufficit etiam pro medio Syllogistico, ut extrema unita in medio sic uno, sine fallacia aequivocationis, concludantur inter se uniri. (N.B. The Vivès as well as the older editions have "unum" for "uniri." Ockham fortunately quotes the text in its more original and intelligible form.)

formal perfection; in the minor, an extrinsic perfection. For law is divine only by reason of its source, the divine Lawgiver.

Similarly, because of the analogical character of divine, we may say that the Mosaic law is simultaneously divine and not divine; that is, "The law is not intrinsically and formally divine, but only extrinsically." And so with other "analogical" terms. If the common term "good" can imply either infinite subsistent goodness or finite and participated goodness, one can say without formal contradiction: "God is good and not good", that is, "God is infinite subsistent goodness and not finite participated goodness."

THE "AUGUSTINIAN" POSITION AS SCOTUS CONCEIVED IT

According to the "Augustinian"[19] viewpoint, our transcendental notions, such as being, goodness, truth, etc., are analogous in the sense that they apply primarily (*per prius*) to God and only secondarily and with qualification (*per posterius*) to creatures.[20] Thus God alone is being, goodness, truth in an unqualified sense. Creatures are being, good and true only in a participated and derivative sense. Now, if this Platonic language is translated into Aristotelian terminology, it can only mean that in actual predication of such transcendental terms there are in reality two meanings and consequently two concepts.

The first of these concepts is being in the simple and unqualified sense, to take the primary transcendental notion. Ac-

[19] By "Augustinian" we do not refer to a distinct philosophical system or school of thought based primarily on St. Augustine. For there is no evidence than any such system or school of thought ever existed. For the scholastics of the 13th and 14th centuries only one basic philosophy was considered consistent with facts, that of Aristotle. But certain scholastics substituted elements of neo-Platonism drawn from St. Augustine and Arabian sources for certain solutions of Aristotle. It is these particular doctrines that are referred to as Avicennian or Augustinian or Averroistic. One such Augustinian element was the theory of illumination. In this connection confer E. Gilson's "Les sources gréco-arabes de l'augustinisme avicennisant" in *Archives d'Histoire doctrinale et littéraire du Moyen Age,* IV (1929), 5-149.

[20] Confer for instance Alexander of Hales, *Summa Theologica,* I, n. 345, t. I, p. 513; n. 347, p. 514; n. 21, p. 32; St. Thomas Aquinas, *Summa Theologica,* I, q. 13, a. 3, c.

cording to Henry of Ghent,[21] it is *ens indeterminatum negative.* This concept is proper to God, not in the sense that it represents God perfectly—it is still a very imperfect concept—but in the sense that it can be predicated exclusively of God. It is only analogous to the concept predicable of creatures (*ens indeterminatum privative,* according to Henry). Furthermore, this concept proper to God is a simple concept. It cannot be analyzed or resolved into two more simple concepts, one expressing, for instance, a common univocal element found in both God and creatures, the other the mode of indeterminability or fulness or infinity. It is an absolutely simple notion of being as incapable of further determination. It is simply being. These two concepts (*ens indeterminatum privative* and *ens indeterminatum negative*) are so similar that we tend to regard them as a single analogical concept.[22]

---

[21] Henry of Ghent, *Summa,* art. 21, q. 2, nn. 6ff; t. II, pp. 315-317. Scotus, *Oxon.* 1, d. 3, q. 2, n. 3; IX, 13b-14a.

[22] Henry of Ghent *Summa,* a. 21, q. 2, n. 6; II, 315: Et ideo absolute dicendum quod esse non est aliquid commune reale in quo Deus communicet cum creaturis, et ita si ens aut esse praedicatur de Deo et creaturis, hoc est sola nominis communitate, nulla rei, et ita non univoce per definitionem univocorum, nec tamen pure aequivoce, secundum definitionem aequivocorum casu, sed medio modo ut analogice. *Ibid.* n. 14; 316-317: Omnis ergo conceptus realis quo aliquid rei concipitur concipiendo esse simpliciter, aut est conceptus rei quae Dei est, aut quae creaturae est, non alicuius communis ad utrumque. Videtur tamen hoc non potentibus distinguere multiplicitatem entis et esse creatoris ab esse creaturae, sicut nec potuit Plato ponens ens esse genus, tanquam sit nominis entis unum aliquid commune conceptum, quod non videtur subtilioribus potentibus distinguere ens, et ejus significata discernere qualis erat Arist. Quod autem nomine entis videatur concipi aliquid commune, est quia, sive concipiatur aliquid, quod est res divina, sive quod est creatura, tamen cum concipitur esse absque eo quod determinate et distincte concipitur esse Dei, vel creaturae, illud non concipitur nisi indeterminate, scilicet non determinando intellectum ad esse Dei, vel esse creaturae. Et habendo respectum ad distinctum intellectum Dei aut creaturae, intellexit Avicenna, si bene intellexit, quod intellectus entis prior est intellectu Dei aut creaturae. Intelligendum tamen quod illa indeterminatio alia est respectu esse Dei, et alia respectu esse creaturae, quia duplex est indeterminatio, una negative, altera vero

The position of Henry of Ghent, as Scotus conceived and described it, may be indicated in diagrammatic form as follows.

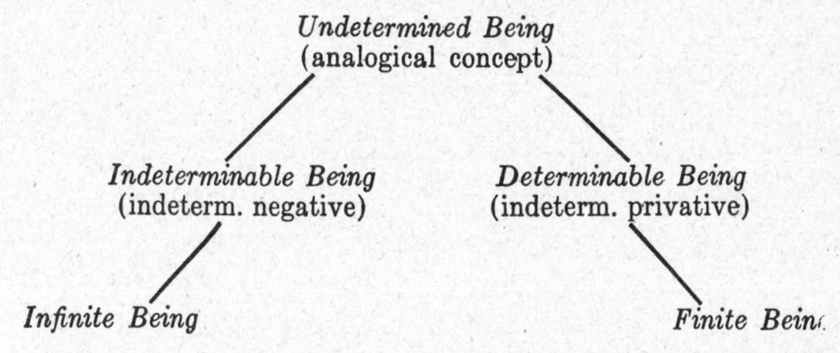

The first concept (*undetermined being*) abstracts as far as possible from all determination. This concept is analogously predicable of God and creatures in the sense that it applies primarily (*per prius*) to that Being which is indetermined because He lacks all potentiality to determination (*indeterminable being.*) Only secondarily (*per posterius*) does it apply to being as indifferent to further determination, that is, to determinable being. This " analogous " concept then contains two almost indistinguishable meanings, one proper to God, the other proper to creatures.[23] The first, *indeterminable being,* negates all deter-

privative dicta. Est enim negativa indeterminatio quando indeterminatum non est natum determinari ad modum, quo Deus dicitur esse infinitus, quia non est natus finiri. Est autem privativa indeterminatio, quando indeterminatum natum est determinari, ad modum quo punctus dicitur esse infinitus, cum non est determinatus lineis, quibus natus est determinari.

[23] *Ibid.* n. 17; p. 317: Per hunc ergo modum, esse indeterminatum per abnegationem convenit Deo, et per privationem creaturae; et quia indeterminatio per abnegationem et per privationem propinquae sunt, quia ambae tollunt determinationem, una tantum secundum actum, alia secundum actum simul et potentiam, ideo non potentes distinguere inter huiusmodi diversa, pro eodem concipiunt esse simpliciter et esse indeterminatum sive uno modo, sive altero, sive sit Dei, sive ut creatura. Natura enim est intellectus non potentis distinguere ea quae propinqua sunt, concipere ipsa ut unum, quae tamen in rei veritate non faciunt unum conceptum. Et ideo est error in illius conceptu.

mination both actual and potential, hence it is called "negatively indetermined being." This concept proper to God is a simple concept. It contains implicitly the notion of illimited or infinite being. The concept, infinite being, merely expresses explicitly the content of the analogous notion of indeterminate being as applied to God. The second notion, *determinable being,* abstracts from actual determination only. It is a common notion in the sense that it is predicable of all finite or actually determined beings, but it prescinds from their actual differences or determinations. Because this notion is still determinable, it contains implicitly the notion of finiteness. This intrinsic mode is expressed explicitly in the concept, finite being.[24]

THE PROBLEM OF DUNS SCOTUS

There is no doubt that God and creatures differ radically from one another, for all of God's perfections are formally infinite while those of creatures are finite.[25] Furthermore, any term that signifies the proper perfection of God (that is, the perfection as infinite) can only be applied analogously (from the metaphysical viewpoint) or equivocally (from the logical standpoint) to the

[24] The fundamental difficulty of Henry's position, as Scotus saw it, is that every concept we apply to God is proper to Him and only analogous to the concepts of creatures through which we somehow come to this proper notion of God. *Rep. Par.* 1, d. 3, q. 1, n. 4; XXII, 94. For Henry the human mind is able to *see through* creatures as it were to what lies beyond, namely God. He uses an analogy. If a wolf were miraculously transformed so that so far as all external sense qualities were concerned it would seem to be a lamb, another lamb by reason of its *vis aestimativa* would perceive the dangerous character of the wolf, nevertheless. Sicut enim aestimativa in brutis, sub intentionibus sensatis subfodiendo, cognoscit intentiones non sensatas, ut nocivi et proficui, sic intellectus sub specie creaturae, quae non repraesentat nisi creaturam, ex suo acumine subfodit ad cognoscendum ea quae sunt et dicuntur de Deo, per speciem alienam ex creaturis, et hoc omnibus tribus modis praedictis. *Oxon. ibid.* p. 14a.

[25] *Oxon.* 1, d. 8, q. 3, n. 16; IX, 596a: Quidquid est in Deo perfectio essentialis, est formaliter infinitum, in creatura vero finitum. *Ibid.* 597a: Igitur cum in Deo quaecumque realitas essentialis sit formaliter infinita, nulla est a qua formaliter possit accipi ratio generis.

perfections found in creatures. Or vice versa, any term that signifies the proper perfection of a creature can only be applied analogously or equivocally to the perfections of God. But the problem is not one regarding a term or name, it is one regarding the concept (signification or meaning attached to the name). *Where do we get this notion which applies properly to God and is only analogous to the concept we apply to creatures?* Where do we get the notion of being, for instance, as absolutely indetermined and hence implicitly including the mode of infinity and therefore proper to God? Once we have it, the theory of analogy follows logically enough. But analogical knowledge is always a relative and comparative knowledge. It is secondary knowledge in the sense that it requires previous knowledge of the terms to be compared, just as judgment is a secondary operation of the human mind because it presupposes conceptual knowledge.

Several possibilities are open by which one might arrive at this knowledge of the other term of comparison, God. They will depend, however, on whether this concept, which includes at least implicitly the modality of infinity, is a simple concept or a composite concept, whether it results from a simple apprehension or abstraction or is attained through an inferential or reasoning process.

According to the Augustinian position as represented by, let us say, Alexander of Hales, St. Bonaventure, and Henry of Ghent, this notion which applies properly or *per prius* to God is an irreducibly simple concept (*simpliciter simplex*). But where do we get such a simple proper notion of God which applies to creatures only in a secondary and analogous sense? Theoretically speaking, several possibilities are open. For example, God could infuse it, says Scotus, either directly (innatism) or by somehow elevating our natural powers (illuminationism).[26] In St. Augustine, Alexander of Hales, St. Bonaventure, Robert Marston, and others we find some form of illumination. In the

[26] *Rep. Par.* 1, d. 3, q. 1, n. 8; XXII, 95b: Dico igitur ad quaestionem quod in intellectu viatoris absolute est possibilis conceptus Dei proprius, quia hoc posset Deus creare infundendo lumen proportionale....; talem tamen conceptum non potest intellectus noster acquirere sua cognitione naturali.

formula of Avicenna, *res et ens sunt primae impressiones animae.*[27] And so too with other transcendental notions like true, good, etc.[28] Another alternative, more naive yet equally effective, is that found in Plato's dialogues. The human soul, according to this doctrine, acquired its knowledge of the absolute Good, the True, and the Beautiful before being incarcerated in the body. The participated likenesses of those archetypal ideas in the visible world recall to mind this knowledge previously acquired. Comparing the real world with the ideal, the philosopher can recognize the analogical character of the two.

But what about the Aristotelian? The natural way of acquiring simple concepts is by means of a phantasm and the activity of the agent intellect. As a result, all our simple concepts represent abstractions of intelligible notes found in sensible quiddities. But what material or sensible quiddity contains even implicitly the mode of infinity? Note, the question is not whether we can reason to a notion which contains either implicitly or explicitly a modality proper to God, but whether we can abstract such a notion. We do not reason to the existence of simple notions but to that of composite notions. A simple notion is that which

---

[27] Avicenna, *Metaph.* tr. 1, c. 6: Dicemus igitur, quod *ens* et *res* et *necesse* talia sunt, quod statim imprimuntur in anima *prima impressione,* quae non acquiritur ex aliis notioribus se... (quoted from the editors of *Opera Omnia S. Bonaventurae*) (Quaracchi, Coll. S. Bonaventurae, 1882-1902) V, 308, note 8.

Confer for example Alexander of Hales, *Summa theologica* I, 72 (Quaracchi edition, 1924) I, 113: "Dicendum quod cum sit 'ens' primum intelligibile, ejus intentio apud intellectum est nota; primae ergo determinationes entis sunt primae *impressiones* apud intellectum: eae sunt unum, verum, bonum, sicut patebit." And in *Op. cit.* I, 24 (I, 36b) he speaks of the "habitum infusum."

[28] Alexander of Hales, *op. cit.* I, 72 (I, p. 113); Robert Marston, *Quaestiones disputatae de Anima* (Quaracchi, Coll. S. Bonaventurae, 1932), q. 3, p. 258: Firmiter teneo unam esse lucem increatam, in qua omnia vera certitudinaliter visa conspicimus. Et hanc lucem credo quod Philosophus vocavit intellectum agentem...quod sit substantia separata per essentiam ab intellectu possibili. . ." ; S. Bonaventura, *Itinerarium Mentis in Deum,* c. 5, n. 3 (*Opera Omnia,* V, 308-9); St. Augustine, *De Videndo Deo,* c. 17 (PL 33, 615-6).

results from a simple act of apprehension. But if we could abstract such a simple proper notion, we would possess a real though imperfect knowledge of God, without recourse to reason—a notion only analogous to that proper to creatures. It is precisely because such an effect exceeds the natural causality of the material object and intellect that a special illumination had been postulated by the " Augustinians ". With them St. Augustine's dictum was more than just a beautiful metaphor: *Bonum hoc et bonum illud, tolle hoc et illud et v i d e Ipsum Bonum, si potes, ita Deum videbis.*[29] From the viewpoint of an illuminationist, one might ask: How is it possible to prescind from the mode of infinity when we predicate being or wisdom or goodness of God? For the Aristotelian, however, the problem is different. How do we know that any being is infinite? How do we know that the mode of infinity is associated with wisdom or goodness to begin with?

We have no simple concepts, says Scotus, that are proper to God.[30] Any concept which contains a modality proper to God (whether explicitly or implicitly) is not the result of simple

---

[29] *De Trinitate* 8, c. 3 (PL 42, 949).

[30] *Rep. Par.* 1, d. 3, q. 1, n. 10; XXII, 96bf: Respondeo, quod loquendo de conceptu simpliciter simplici, nullum conceptum proprium de Deo possumus naturaliter cognoscere, sed solum talem conceptum communem sibi et creaturis, in quo a creaturis non distinguitur. Sed quia possumus multos tales conceptus communes Deo et creaturis abstractos concipere, unum etiam illorum per alium determinare, ideo sic possumus abstrahere a creaturis, et ipsos sic cognoscere Deum secundum aliquem conceptum sibi proprium, non simpliciter simplicem, sed secundum conceptum resolubilem in duos conceptus simpliciter simplices. Verbi gratia, a creatura possumus abstrahere conceptum *bonitatis* vel *boni,* similiter conceptum *summi,* cum non sit processus in infinitum, et ita componendo apprehendo conceptum *boni summi,* quod est proprium Dei, et convenit tantum Deo. Sic etiam de *ente infinito,* quamvis enim uterque conceptus simpliciter simplex sit communior conceptu Dei, conveniens univoce Deo et creaturis, tamen post determinationem uterque conceptus particularisatur, et fit conceptus proprius Deo, sic quod solum illi convenit. Et hoc est possibile, quia secundum Philosophum 2. Posterior. c. 8. *licet quaelibet pars definitionis sit in plus suo definito, totum tamen in aeque.* Confer also, *Oxon.* 1, d. 3, q. 2, nn. 8-9; IX, 19-20; *Collationes* XIII, n. 4; V, 202a; *Ibid.* n. 5; 202b; *Rep. Par.* 1, d. 3, q. 2, n. 2; XXII, 98b, etc.

apprehension but of a synthetic activity of the mind. Such notions as Pure Act, First Cause, Unmoved Mover, Being as infinite, etc. are all composite notions. Before such notions can be called real, it must be demonstrated that the two or more simple notes must coexist in one and the same subject. In other words, the conclusions must be established that "Some cause is first," "Some mover is unmoved," "Some being is not finite (infinite)," etc.[31]

A proposition like the last is valid only if some reality exists of which both being and not-finite can be predicated. Such a being is, by definition, God. But this notion of being predicable of God which appears in the conclusion of the syllogism is the same notion, so far as conceptual content goes, as that which appeared in the premises and ultimately was abstracted from sensible quiddities. For terms in the conclusion cannot take on any additional meaning over and above what they have in the premises. If the meaning of the term is the same in premise and conclusion, by Scotus' definition it is univocal. In a word, we have a concept univocally predicable of God and creatures. Note, it is an imperfect concept because it does not signify the intrinsic mode of either God or creatures but merely a common aspect to be found in both. But this imperfection of the concept destroys neither its validity nor its reality. This common univocal concept becomes a proper concept of God only by the addition of some other note, such as "first," "supreme," or "infinite." All our proper notions, therefore, are composite and arrived at by way of inference ("*conceptus conclusi per modum complexionis*").[32] And in every proper concept of God there is at least

[31] That is why Scotus in formulating his proofs for the existence of God does not ask whether an Infinite Being exists, but "Utrum in entibus sit aliquid actu existens infinitum?" *Oxon.* 1, d. 2, q. 1; VIII, 393b. Because notions like "First Cause," "Unmoved Mover," "Infinite Being" are arrived at by way of demonstration, Scotus calls them "conceptus conclusi per modum complexionis." Confer *Coll.* XIII, n. 4; V, 202a.

[32] *Coll.* XIII, n. 4, V, 202ab: Dico quod conceptus conclusi per modum complexionis conveniunt Deo, nec conveniunt creaturae; hujusmodi sunt conceptus compositi, non autem simplices conceptus, cujusmodi sunt conceptus entis, boni, etc. nam tales conceptus dicuntur univoce de Deo et creatura.

one simpler common concept predicable univocally of God and creatures. Scotus says, at least one element of the composite concept is common, namely, the subject of the conclusion of the demonstration. For the conclusion may be either an affirmative or a negative proposition, for example, " Some being is first," and " Some cause is not caused." In either case, however, the subject of the conclusion must contain some positive element common to God and creatures.[33]

Once we possess such a proper notion, particularly one which includes the modes of infinity, we may arbitrarily designate it by the same name as we use for the simpler common concept. *Sensus verborum est ad placitum.* Scotus, we believe, would have no quarrel with St. Thomas when the latter says: " When we say that God is good... the meaning is, Whatever good we attribute to creatures pre-exists in God, and in a more excellent and higher way." [34] Such a concept, like the perfection of God which it signifies, is only analogous to what we mean by " good " as proper to creatures. Or we may use Scotus' own example regarding the term " being." In answer to the question: Is being predicated univocally of all things? he writes.

> As to the question, I grant that being is not predicated univocally of all things. Neither is it predicated equivocally, for something is said to be predicated equivocally when those things of which it is predicated are not attributed to one another. But when they are attributed, then [it is predicated] analogically. But because it [i. e. the term being] does not have one concept [corresponding to it], it signifies all things essentially according to their proper perfection and simply equivocally according to the logician. But because those things which are signified are essentially

[33] Purely negative knowledge of God would be no knowledge at all. Confer *Oxon.* 1, d. 3, q. 2, n. 2; IX, 10a: Et quantumcumque procederetur in negationibus, vel non intelligeretur Deus magis quam nihil, vel stabitur in aliquo conceptu affirmativo, qui est primus.

[34] *Summa Theol.* I, q. 13, a. 2, c.: Cum igitur dicitur: Deus est bonus, non est sensus: Deus est causa bonitatis, vel: Deus non est malus: sed est sensus: id quod bonitatem dicimus in creaturis, praeexistit in Deo, et hoc quidem secundum modum altiorem.

attributed to one another, [being is predicated] analogously, according to the metaphysician.[35]

---

[35] In the light of the different interpretation we have given to Scotus' theory of univocity, this text takes on new interest. In the Vivès edition, it reads as follows: "Ad quaestionem, concedo quod ens non dicatur univoce de omnibus entibus, non tamen aequivoce, quia aequivoce dicitur aliquid de multis, quando illa, de quibus dicitur non habent attributionem ad invicem, sed quando attribuuntur, tunc analogice. Quia ergo non habet conceptum unum, ideo significat omnia essentialiter secundum propriam rationem, et simpliciter aequivoce secundum Logicum; quia autem illa quae significantur inter se essentialiter attribuuntur, ideo analogice secundum Metaphysicum realem." (*Metaph.* 4, q. 1, n. 12; VII, 153a)

The Scotistic Commission, in a private communication to the author, states that of the 17 manuscripts which they have examined, 12 (among which are some of the best and oldest manuscripts) have the reading of the Vivès edition, but that the 5 remaining manuscripts omit the "Ad quaestionem" and read simply, "Concedo quod ens non dicatur aequivoce, etc...", the remainder of the passage is essentially the same. Though a final decision has not been reached, the members of the Commission are, at the present time, inclined to regard the reading of the 5 manuscripts which delete the phrase "ens non dicatur univoce de omnibus entibus" as the better reading, noting that the age of the manuscripts is not necessarily a help in determining the genuine reading. The preference for the latter reading may have been occasioned by the fact that it does away with the phrase, which, in the words of Gilson, "a fait le désespoir de générations de scotistes, fermes partisans de l'univocité de l'être, qui s'y trouvent devant une négation radicale de l'univocité et une décision expresse de Duns Scot en faveur de l'analogie. ("Avicenne et le point de départ de D. Scot" *Archives d'HDLMA,* II (1927) 105).

However, the deletion of this passage, "Concedo quod ens non dicatur univoce de omnibus entibus...", does not solve the difficulty to any considerable extent, since Scotus goes on to assert that being is predicated analogously or equivocally (and hence, not univocally), an assertion which is clearly found in the logical works, for instance, *Super Praed.* q. 4, n. 7; I, 447b: "Propter hoc dicendum quod hoc nomen *ens,* simpliciter est aequivocum primo modo aequivocationis, ad haec decem Genera prima, praecipue propter hoc ultimum; quia certum est substantiam significari sub propria ratione, et accidens alio modo; quia si sub ratione propria significetur accidens per hoc nomen *ens,* hoc esset sibi proprium sub ratione, qua attribuitur substantiae, vel sub aliqua alia consimili, illa est propria ratio accidentis: adhuc sequitur utrumque significare sub propria ratione." Hence the reading one adopts makes little difference.

Scotus in short does not deny that the proper perfections of God are conceived in a concept whose content is only analogous to that of the concept which designates the perfections proper

---

The interpretation we have adopted, however, obviates the whole difficulty very simply. The statement " Concedo quod ens non dicatur univoce de omnibus entibus, etc." does not involve a " direct and explicit negation of a univocal concept of being and the univocal predication of being ", as Shircel (*Univocity of the Concept of Being in the Philosophy of John Duns Scotus,* Washington, D. C., Catholic University of America Press, 1942, p. 108) and Gilson (*loc. cit.*) maintain. In fact, Scotus is not speaking of the *concept* of being at all, but of the *term,* " being." He says explicitly in the *Super Praed.* " hoc *nomen* ens " and the context in the *Metaphysics* also indicates clearly that he is speaking of the term for he speaks of it as singular. Yet he indicates that corresponding to being we do not have a single concept but a plurality of concepts. Therefore, he argues, it (namely, the *term* " being ") can signify all things (God, creatures, substance, accident) *secundum propriam rationem.* The point to emphasize here is that Scotus is speaking of *proper concepts,* and not of a *common* concept. It is precisely because being " utrumque significat sub propria ratione " that it is " aequivocum Substantiae et Accidenti " (*Super Praed.* n. 6, I, 447b). Scotus clearly grants that the term " being " can be applied to proper concepts, i.e. concepts which include not only a *common ratio* but the intrinsic mode essentially and formally identical with the perfection in question. Since Scotus is obviously speaking of the use of the term " being " to signify *properly* and hence to have corresponding to it a *proper concept,* it is equally patent that the term is not being predicated univocally, but equivocally, or analogously, if you will. But this does not imply that we cannot abstract from the proper modalities to form a common univocal, albeit imperfect, concept. Neither does it deny that the term " being " can be used to signify only the common aspect of things (*Super Elench.* q. 15, n. 7, II, 22b), in which case it is predicated *in quid* and univocally, as we shall show in the following chapter.

That Scotus himself did not consider his statement in the *Metaphysics* a denial of the possibility of forming such a univocal concept of being is evident on this score. In a previous question of the same work, in answer to the objection that being as object of the intellect would imply the possibility of having a univocal concept of being, he concedes the implication and refers to this question where the matter is touched on in connection with the argument for univocation based on the possibility of being certain that something is a being, while doubting whether it is substance or accident. *Meta.* 2, q. 3, n. 22; VII, 112a: " Ad primum aliter dicitur, quod ens est unius rationis, ut tangitur in illa quaestione 4 hujus, alioquin non esset certum quod aliquod est ens, et dubium si

to creatures. The same is true of substance and accident when the proper notions of each are designated by a common name, say "being." The important thing is to remember how we arrived at such a proper concept of God or of substance. For the Aristotelian is must be through the medium of common and univocal notions—notions which for all their imperfection, are as real and as valid as the proper composite notions of which they are the elements.

In defense of his position, Scotus advanced three arguments. The first is to show that the common concepts we predicate of God and creatures actually prescind from the intrinsic modes entirely; the second, that a simple concept proper to God is impossible in the Aristotelian theory of ideogenesis; the third, that the famous threefold way of arriving at a knowledge of God (*via affirmationis, negationis et eminentiae*) implies that a common univocal concept exists.

ARGUMENTS FOR UNIVOCATION

First argument:

A very simple test to discover whether or not our common notions of being or of any pure perfection have been completely divorced from the intrinsic mode is to apply to the concept in question contradictory predicates. If being, for instance, implicitly includes in its very concept the mode of infinity when applied to God, then the meaning of the term "being" as applied to God and the meaning of the term "infinite being" are form-

---

substantia, vel accidens." It is interesting to note that throughout the whole argumentation in this fourth question, Scotus is careful to defend the validity of this argument for univocity.

Scotus, in short, saw no contradiction in admitting that the term "being" can be predicated either equivocally (analogously) or univocally depending upon whether it signifies all things *properly* or not. In the first instance, it signifies the intrinsic mode as well as the common *ratio* of being and hence has as many corresponding proper concepts as there are intrinsically different classes of beings. In the second instance, it signifies things only imperfectly, since the mind abstracts from the proper intrinsic modes. But in this case we have not several concepts, but one *univocal common* concept.

ally identical. Hence I cannot simultaneously affirm and deny contradictory predicates of this notion.

But if the meaning of "being" as applied to God and of "infinite being" is formally identical, how is it that I can be both certain and doubtful about one and the same notion? Yet, Scotus argues, I can be certain that God is a being and still in doubt whether He is infinite or not.[36] I may know that He is wise, without knowing that He is infinite self-subsistent Wisdom. I may define God as the Cause of the visible universe and yet not know whether He is the Uncaused Cause. The child at its mother's knee learns that God is good, but who will say that the content of its childish concept includes even implicitly the modality of infinity?

The same argument Scotus applies to substance and accident.[37] Being can be predicated univocally of both and is so predicated when the concept corresponding to the term expresses merely their common elements. The example Scotus uses might have been taken from a modern book on physics. We are quite certain, he says, that light is a being. But we are still in doubt

---

[36] *Oxon.* 1, d. 3, q. 2, n. 6; IX, 18a: Primo sic, omnis intellectus certus de uno conceptu, et dubius de diversis habet conceptum de quo est. certus, alium a conceptibus de quibus est dubius, subjectum includit praedicatum; sed et intellectus viatoris potest esse certus de aliquo quod sit ens, dubitando de ente finito vel infinito, creato vel increato; ergo conceptus entis de aliquo est alius a conceptu isto vel illo et ita neuter ex se sed in utroque illorum includitur; ergo univocus.—*Oxon.* 1, d. 3, q. 3, n. 9; IX, 109; *Collatio* XIII, n. 3; V, 201b; *Collationes in editione Waddingi non inclusae*, q. 3 (ed. by C. R. S. Harris, *Duns Scotus* [Oxford, Clarendon Press, 1927] II, 371): Si respondeatur quod ille conceptus qui dicitur unus, non est unus simpliciter, sed secundum quid, scilicet aggregatione; quia continet duos conceptus, licet non percipiatur distincte ab intellectu. Contra: si sunt duo conceptus ille unus de quo est certus, jam non esset certus de uno conceptu, sed de duobus; et ita esset certus de uno conceptu et non certus.

[37] *Metaphys.* 4, q. 1, n. 6; VII, 148b: Experimur in nobis ipsis, quod possumus concipere ens, non concipiendo hoc ens in se vel in alio, quia dubitatio est quando concipere ens utrum sit ens in se vel in alio, sicut patet de lumine, utrum sit forma substantialis per se subsistens, vel accidentalis existens in alio sicut forma. (Note: This argument is among those which are first refuted but it is afterwards reinstated in n. 13; 154b-155a.)

whether light is fundamentally a substance or only an accidental modification of a substance. Modern physicists tell us that light may be considered as a " wave and/or a particle." They have only altered the formula; they have not resolved the doubt. Yet we are all quite certain that light is something. And further knowledge will not change or alter this basic conception. It can only enrich this concept from without.

If the objection of Henry of Ghent [38] is raised that being is in reality a term with two meanings but so closely related, so extremely similar that we cannot separate them even in concept, then, says Scotus, we might as well admit that all univocal knowledge is impossible.[39] For we certainly have no more cogent way of demonstrating a formal unity of concept than by means of the principle of contradiction. We might just as well object that *man* is not predicated univocally of Plato and Socrates, but that each possesses a different kind of human nature, so similar, however, that we cannot distinguish the one from the other. And, of course, if univocal knowledge is impossible, the basis of the principles of identity, contradiction and excluded middle is destroyed and sheer scepticism will result.

Second argument:

In the present life, every simple concept caused naturally must be produced by that which naturally motivates the human intellect. According to Aristotle, the motivating factors are the agent intellect together with the phantasm or the object revealed therein. Now any simple concept that is not univocal but different from and only analogous to the object revealed in the phantasm, could not arise through the activity of the naturally

---

[38] Henry of Ghent, *Summa,* a. 21, q. 2, n. 17; t. II, 317: Natura enim est intellectus non potentis distinguere ea quae propinqua sunt concipere ipsa ut unum, quae tamen in rei veritate non faciunt unum conceptum. Et ideo est error in illius conceptu.

[39] *Oxon.* 1, d. 3, q. 2, n. 7; IX, 18b-19a: Ex ista evasione videretur destructa omnis via probandi unitatem alicujus conceptus univocam. Si enim dicis, hominem habere unum conceptum ad Socratem et Platonem, negabitur et dicetur quod sunt duo, sed videntur unus propter magnam similitudinem.

motivating factors mentioned above. But if even our simplest notions of God are analogous to those through which creatures are known, all natural knowledge of God would be impossible.

The object revealed in the phantasm cannot produce a simple proper concept of itself and a second simple proper notion of something else, unless the notion of the second thing be contained either essentially or virtually in the first. A baseball, for instance, could produce a simple proper notion of itself as a sphere and also a simple proper notion of a circle, for the notion of circularity is virtually contained in the notion of sphericity. But it could not give rise to a simple notion of triangle or pentagon. No created object, however, contains God either essentially or virtually. Consequently, no created object can produce a simple notion that is proper to God and only analogous to any essential or virtual perfection it may contain. The only other way in which we might arrive at a proper knowledge of some other object would be by way of discursive reasoning. But such reasoning presupposes certain simple univocal elements and results, not in a simple, but in a composite real notion predicable exclusively of God.[40]

---

[40] *Ibid.* nn. 8-9; 19-20: Secundo principaliter arguo sic: nullus conceptus realis causatur in intellectu viatoris naturaliter, nisi ab his quae sunt naturaliter motiva intellectus nostri, sed illa sunt phantasma vel objectum relucens in phantasmate, et intellectus agens. Ergo nullus conceptus simplex fit modo naturaliter in intellectu nostro, nisi qui potest fieri virtute istorum. Sed conceptus, qui non esset univocus alicui objecto relucenti in phantasmate, sed omnino alius et prior, ad quem iste haberet analogiam, non posset fieri virtute intellectus agentis et phantasmatis, ut probabo; ergo talis conceptus alius analogus, quo ponitur naturaliter in intellectu viatoris nunquam erit et ita non poterit naturaliter haberi aliquis conceptus de Deo, quod est falsum. Probatio assumpti, objectum quodcumque, sive relucens in phantasmate, sive in specie intelligibili cum intellectu agente vel possibili cooperante, secundum suum proprium, et conceptum ultimum suae virtutis facit in intellectu, sicut effectum sibi adaequatum, conceptum omnium essentialiter vel virtualiter inclusorum in eo; sed ille alius conceptus qui ponitur analogus, non est essentialiter vel virtualiter inclusus in isto, nec est iste. Ergo ille non fiet ab aliquo tale movente. Et confirmatur ratio, quia praeter conceptum suum proprium adaequatum et inclusa in ipso altero praedictorum duorum modorum, nihil potest cognosci ex isto objecto nisi per discursum, sed

The same argument can be applied to the nature of our notion of substance. For substance is known to the intellect only through the medium of accidents. The intellect has no immediate contact with substances existing in the extra-mental world. That is quite clear, says Scotus, from the instance of the consecrated Host. If we were immediately aware of substance we would likewise know when it is absent, just as we recognize darkness, which is the absence of light. But we only know through faith that the substance of bread is no longer present in the consecrated Host.[41]

---

discursus praesupponit cognitionem istius simplicis ad quod discurritur. Formetur ergo breviter ratio sic, videlicet nullum objectum facit conceptum simplicem et proprium sui in aliquo intellectu, et conceptum simplicem et proprium alterius objecti, nisi contineat illud aliud objectum essentialiter vel virtualiter; objectum autem creatum non continet increatum essentialiter vel virtualiter, ergo, etc. Secunda pars minoris, scilicet quod objectum creatum non contineat virtualiter increatum et hoc sub ea ratione sub qua sibi attribuitur, ut posterius essentialiter attribuitur priori essentialiter, probatur, quia est contra rationem posterioris essentialiter, includere virtualiter suum prius, et etiam patet quod non continet essentialiter increatum, secundum aliquid omnino sibi proprium et non commune; ergo non facit conceptum simplicem et proprium enti increato.

[41] *Oxon.* 1, d. 3, q. 3, n. 9; IX, 109a-b: Secundam rationem pertracto sic, sicut est argutum, quod Deus non est a nobis cognoscibilis naturaliter, nisi ens sit univocum creato et increato, ita potest argui de substantia et accidente; cum enim substantia non immutet immediate intellectum nostrum ad aliquam intellectionem sui, sed tantum accidens sensibile, sequitur, quod nullum conceptum quidditativum habere poterimus de ea nisi sit aliquis talis, qui possit abstrahi a conceptu accidentis; sed nullus talis quidditativus, abstrahibilis est a conceptu accidentis nisi conceptus entis, ergo, etc. Hoc autem quod suppositum est in ratione ista de substantia, quod non immutat intellectum nostrum immediate ad actum circa se, probatur sic: quia quidquid praesentia sua immutat intellectum, absentia illius potest naturaliter cognosci ab intellectu, quando non immutatur sicut apparet *secundo de Anima,* quod visus est tenebrae perceptivus, quando scilicet lux non est praesens, et ideo tunc visus non immutatur a substantia; igitur si intellectus naturaliter immutatur a substantia immediate ad actum circa ipsam, sequeretur quod quando substantia non esset praesens, posset naturaliter cognosci non esse praesens, et ita naturaliter posset cognosci in hostia altaris consecrata non esse substantiam panis, quod est manifeste falsum. Nullus igitur conceptus

What the mind actually contacts is the sensible phantasm, which materially contains the quiddity of the accident. Our concept of substance, like our concept of God, is a composite notion, a mental construct built around the core of an ultimate univocal element—the notion of being as common to both substance and accident. To deny the possibility of a univocal concept is to destroy the bridge by which we contact substance and is equivalent to declaring that the essences of things are unknown and unknowable.[42]

Third argument:

Scotus' third argument is similar to the above. It involves an analysis of the threefold way of knowing God attributed to Pseudo-Dionysius.[43]

Every metaphysical inquiry about God, says Scotus, proceeds as follows: We consider the formal perfection of a thing, and then remove from it any imperfection it possesses by reason of its mode of existence in creatures. Then finally to this formal notion we add the mode of infinite perfection and attribute it properly to God. Take, for instance, the idea of wisdom, intelligence, or free will. Any imperfections associated with these perfections as they exist in creatures must be removed, for instance, the idea of discursive reasoning characteristic of man's intellect, the fact that wisdom is a quality accidental to man, etc. We retain, however, the sheer formal notion of wisdom or intelligence, which as such says nothing of either finite or infinite.

---

quidditativus habetur naturaliter de substantia immediate causatus a substantia, sed tantum causatus vel abstractus primo de accidente, et illud non est nisi conceptus entis.

[42] *Ibid.* n. 11; 110b: Ita nihil cognosceretur de partibus essentialibus substantiae, nisi ens sit commune univocum eis et accidentibus.

[43] *Oxon.* 1, d. 8, q. 3, n. 2; IX, 581b: Ad hoc [namely, in favor of the position of Henry of Ghent who denies any univocal common concept] adducitur intentio Dionysii *de Divinis nominibus,* qui ponit tres gradus cognoscendi Deum, scilicet per eminentiam, per causalitatem, et per abnegationem; et ponit illam cognitionem per abnegationem esse ultimam, quando removentur a Deo omnia illa, quae sunt communia creaturis; ergo non intelligit ipse, quod aliquis conceptus, qui est abstractus a creaturis, remaneat in Deo, secundum quod fuit communis creaturae.

To this formal *ratio* we add the mode of infinite perfection, and arrive at a concept proper to God. But the whole procedure is invalid—as Kant pointed out later—unless the common element, the sheer formal *ratio,* can be validly predicated of God, and that independently of the addition of the mode of infinity.[44]

In another passage, Scotus describes in more detail how the human intellect by its natural powers rises to a knowledge of God. The more universal intelligible species can be abstracted from the less universal. Thus the notions of " living being " and " animal " are contained in the less universal notion, " man ". Created beings or objects are not only capable of producing in the intellect a notion proper to themselves but of more universal and common notions.

> And thus creatures which impress their proper [intelligible] *species* in the intellect can also impress the *species* of the transcendentals that are common to themselves and to God. And then the intellect by its own power can use many [intelligible] species simultaneously to conceive these [transcendentals] together, for instance, the species of *good,* the species of *highest,* the species of *act,* in order to conceive the *Highest Good,* which is *all act.*[45]

---

[44] *Ibid.* d. 3, q. 2, n. 10; IX, 20b-21a: Tertio sic: omnis inquisitio metaphysica de Deo procedit sic, scilicet considerando formalem rationem alicujus, et auferendo ab illa ratione formali imperfectionem quam habet in creaturis, et reservando illam rationem formalem, et attribuendo sibi omnino summam perfectionem, et sic attribuendo illud Deo. Exemplum de formali ratione sapientiae vel intellectus vel voluntatis; consideratur enim primo in se et secundum se, et ex hoc quod ratio non includit formaliter imperfectionem aliquam, nec limitationem, removeantur ab ipsa imperfectiones quae concomitantur eam in creaturis, et reservata eadem ratione sapientiae et voluntatis, attribuuntur ista Deo perfectissime; ergo omnis inquisitio de Deo supponit intellectum habere conceptum eumdem univocum quem accipit ex creaturis.

[45] *Ibid.* n. 18; 36ab: Dico quod ista quae cognoscuntur de Deo, cognoscuntur per species creaturarum, quia sive universalius et minus universale cognoscantur per eamdem speciem minus universalis, sive utrumque habeat speciem intelligibilem sibi propriam, saltem illud quod potest imprimere, vel causare speciem minus universalis in intellectu, potest etiam causare speciem cujuscumque universalioris, et ita creaturae quae imprimunt proprias species in intellectu, possunt etiam imprimere species transcendentium quae communiter conveniunt eis et Deo. Et tunc intel-

Our proper notions of God, then, are built up of simple affirmations and negations of intelligible notes discoverable in creatures. Being, the coextensive attributes, unity, truth, goodness, the other pure perfections, like beauty, wisdom, intelligence, free will, and the like, in fact, the vast majority of our transcendental notions, fall into this class of common univocal elements. Like the tesserae used by the artist, these simple transcendental notions are pieced together to form a mosaic concept of God. While the concept, like the picture, is indeed composite, the reality which it represents is simple and uncomposed.[46]

SUMMARY

As has been noted, we are primarily interested in the doctrine of univocation in so far as it clarifies the general theory of transcendentality of Scotus. By way of summary, then, we call attention to several important distinctions.

The first is between what might be called a transcendental term and a transcendental concept. By transcendental terms we mean those which are used to designate transcendental concepts, such as being, wisdom, truth, and so on. Terms may have more than one meaning; concepts, however, can have but one. Terms like " being," " unity," and " truth," can be used (whether arbitrarily or by custom) to indicate a common aspect possessed by several different things or they may go further and signify also the distinctive mode of existence that the perfection in question possesses in the respective subject. In the first case, since the meaning of the term remains the same in predication, we have a common univocal concept. In the second instance,

---

lectus propria virtute potest uti multis speciebus simul ad concipiendum illa simul, quorum sunt istae species, puta specie boni, specie summi, specie actus, ad concipiendum summum bonum et actualissimum, quod apparet sic per locum a minori. Imaginativa enim potest uti speciebus diversorum sensibilium, ad imaginandum compositum ex illis diversis, sicut apparet imaginando montem aureum.

[46] *Collatio* XIII, n. 5; V, 202b: Ideo licet in altissimo conceptu nihil concipimus, nisi modo composito, tamen intendimus conceptum simpliciter simplicem, quia illum habemus in communi tantum, et confusissimum, et communem Deo et creaturae.

there is no common concept at all but as many distinct, but similar, concepts as there are different subjects. The common term in the latter instance is predicated equivocally (according to the logician) or analogously (according to the metaphysician). The example used by Scotus is that of being. When the term designates everything according to the proper *ratio* of each it is predicated analogously or equivocally. Yet is is possible to prescind from all differentiation and to signify by the term merely a common aspect, in which case both the term and the concept are predicated univocally.

This brings up a second distinction, that which exists between a common and a proper concept of either God or substance. Unless some form of innatism, illuminationism, or ontologism be admitted, all affirmative simple notions of God and substance are common and univocally predicable of creatures and accidents respectively. Such notions are imperfect and cannot be used to differentiate God from creatures or substance from accident. Our proper notions (those which apply exclusively to God or to substance) are all composite notions. Since purely negative knowledge is no knowledge, such composite concepts must contain at least one simple affirmative element. Hence all proper notions presuppose at least one notion univocally common to God and creatures, to substance and accident respectively. In this sense, purely analogical knowledge of God or of substance is naturally impossible. Underlying all analogy is some form of univocation. For this reason Scotus writes:

> There is no greater analogy than that which exists between creatures in reference to God in virtue of their reason of being. Nevertheless, this notion of being which belongs primarily and principally to God, is predicated really and univocally of creatures. And the same is true of goodness, and wisdom, and the like.[47]

[47] *Oxon.* 2, d. 12, q. 2, n. 8; XII, 604a: Nulla enim major est analogia, quam sit creaturae ad Deum in ratione essendi, et tamen sic esse, primo et principaliter convenit Deo, quod tamen realiter et univoce convenit creaturae; simile est de bonitate et sapientia et hujusmodi.

As a final observation in regard to the transcendental concept of *being*, note that every concept common to God and creatures prescinds perfectly from the modalities of finite and infinite, necessary and contingent, and the like. This is essential to Scotus' solution of the paradox of being—a point which will be discussed in the following chapter. Note further in this connection that the common notion of being, a notion which is irreducibly simple, is abstracted from finite or created beings—beings which neither essentially nor virtually contain the mode of infinity. Hence it is impossible that the common *concept* of being should contain virtually or essentially the concepts of infinity or finiteness, etc.[48] This is extremely important for an understanding of the "virtual primacy of being."

[48] Confer in particular the second argument for univocation, note 41.

PART TWO

THE CLASSES OF TRANSCENDENTALS IN PARTICULAR

CHAPTER IV

Transcendental Being

Scotus, like the scholastics in general, considered " being " as the first or primary transcendental.[1] But what did Scotus understand by being as a transcendental? Why did he consider it primary? How is it related to the remaining transcendentals? These questions, particularly the last, are not easily answered. Few conceptions of Scotistic metaphysics have taxed the ingenuity of interpreters more than that of the interrelation of being and its attributes and ultimate differences. The solution suggested in the following pages, it must be emphasized, is only tentative.[2] While it still leaves many other subtle difficulties untouched, particularly in regard to being as the subject of metaphysics, it gives a consistent and coherent picture of Scotus' position. As such it may serve as a basis for further investigation.

Scotus' conception of transcendental being can be understood more readily when analyzed from the standpoint of the genesis

---

[1] *Oxon.* 1, d. 8, q. 3, n. 19; IX, 598b: Non oportet ergo transcendens ut transcendens, dici de quocumque ente, nisi sit convertibile cum primo transcendente, scilicet cum ente.

[2] A treatise that could prove extremely helpful in solving some of the difficulties connected with the problem of being and its attributes is the unfinished tract *De Cognitione Dei* attributed to Scotus. The authenticity of this tract, however, has not yet been established. At the present time only one manuscript is known to contain it (Merton College, MS. Cod. XC, ff. 147r-154v). C. R. S. Harris has edited it in *Duns Scotus* (Oxford, Clarendon Press, 1927) II, pp. 379-398.

of knowledge. The section where Scotus discusses the primacy of being most extensively[3] is simply a further organic development of a problem raised in the previous question, namely: What is first known (*primum cognitum*) by the intellect?

While Scotus discusses three meanings which may be attributed to *primum cognitum*,[4] we are interested only in two interpretations: first object in the order of origin, and first object in the order of adequation. By the former, we mean the content of the first notion that the intellect forms as it passes from a state of ignorance to a state of knowledge. In determining Scotus' position on this point, we learn a great deal about the content of the transcendental notion of being. By " first object in the order of adequation " is meant the primary object of the faculty as such. What object is proper or commensurate to our intellect? The answer to this question sheds light on the relation of being to the other transcendentals.

FIRST OBJECT BY WAY OF ORIGIN

From the standpoint of the origin of man's knowledge, Scotus considers the intellect in its present state.[5] In this connection several points should be borne in mind. Whatever is to be said of man's intellect in a separated state, or even as associated with a glorified body, in the present life it is subject to the following restrictions. First of all, all real knowledge begins with the senses.[6] Even in the formation of such transcendent concepts as those of God and substance, the constant cooperation of the

---

[3] *Oxon.* 1, d. 3, q. 3.

[4] Scotus discusses the meaning of *primum cognitum* in three orders. " Dico quod triplex est ordo intelligibilium in proposito. Unus est ordo originis vel secundum generationem; alius est ordo perfectionis; tertius est ordo adaequationis vel causalitatis praecise." (*Oxon.* 1, d. 3, q. 2, n. 21; IX, 47b.)

[5] *Oxon.* 1, d. 3, q. 2, n. 1; IX, 9a: Juxta hoc quaero: Utrum Deus sit primum cognitum a nobis naturaliter pro statu isto?; *Ibid.* q. 3, n. 5-6; 98aff.

[6] *Metaph.* 6, q. 1, n. 9; VII, 309b: Cognitio nostra oritur a sensu.

phantasy is required.[7] Secondly, our intellect does not immediately contact substances existing in the extra-mental world.[8] Thirdly, singularity *qua* singularity is not directly perceived, but the intellect's first intention terminates primarily at the common nature concretized in singular quiddities.[9]

It goes without saying that the mind does not jump from a state of total ignorance to a state of full and complete scientific knowledge regarding any particular object, say even such a simple quality as redness. The first time the sensation of redness is experienced, the intellect does not recognize the color as one of the several similar hues that make up the color band lying between orange and violet in the physical spectrum. Before

---

[7] *Oxon.* 1, d. 3, q. 3, n. 24; IX, 148-9: Stabilitum est autem illis legibus sapientiae, quod intellectus noster non intelligat pro statu isto, nisi illa quorum species relucent in phantasmate, et hoc sive propter poenam originalis peccati, sive propter naturalem concordiam potentiarum animae in operando, secundum quod videmus quod potentia superior operatur circa idem circa quod inferior, si utraque habeat operationem perfectam, et de facto ita est in nobis quod quodcumque universale intelligimus, ejus singulare actu phantasiamur. Ista tamen concordia, quae est de facto pro statu isto, non est ex natura nostri intellectus, unde intellectus est, nec etiam unde in corpore est, tunc enim in corpore glorioso necessario haberet similem concordiam, quod falsum est. *Ibid.* q. 2, n. 18; 36a-b: Dico, quod ista quae cognoscuntur de Deo, cognoscuntur per species creaturarum, quia sive universalius et minus universale cognoscantur per eamdem speciem minus universalis, sive utrumque habeat speciem intelligibilem sibi propriam, saltem illud quod potest imprimere, vel causare speciem minus universalis in intellectu, potest etiam causare speciem cujuscumque universalioris, et ita creaturae quae imprimunt propriae species in intellectu, possunt etiam imprimere species transcendentium, quae communiter conveniunt eis et Deo. Et tunc intellectus propria virtute potest uti multis speciebus simul ad concipiendum illa simul, quorum sunt istae species, puta specie boni, specie summi, specie actus, ad concipiendum summum bonum et actualissimum, quod apparet sic per locum a minori. Imaginativa enim potest uti speciebus diversorum sensibilium, ad imaginandum compositum ex illis diversis, sicut apparet imaginando montem aureum.

[8] Confer chapter III, p. 52; also *Metaph.* 7, q. 3, n. 3; VII, 362a.

[9] Confer chapter II, p. 29.

the mind is able to define it, or even to begin to say what it is, it passes through the stage of " confused knowledge ".[10]

" Confused knowledge " is understood here in a technical sense. It does not mean that the object is perceived only in a blurred or hazy sort of way, as the eye might perceive the dim, shadowy shapes in a fog.[11] It is opposed rather to what we might call scientific knowledge, or knowledge by way of definition. A teamster, for instance, may know a great deal about horses without being able to give a philosophical or biological definition of the *Equus caballus*. From the philosophical point of view, however, the greater part, if not all, of his knowledge would be considered confused. We know something confusedly (*confuse*), Scotus explains, when our notion stands for the object in much the same way as the name does; we know it distinctly (*distincte*), when we are able to define it.[12] A name, according to Scotus,

---

[10] *Oxon.* 1, d. 3, q. 2, n. 27; IX, 51a: Confuse autem cognoscere est quasi medium inter ignorare et distincte cognoscere.—Confer also Joannes Canonicus, *Quaestiones super Octavos Libros Physicorum*, lib. 1, q. 2. (Venice, Octavianus Scotus, 1481) f.10ra.

[11] " Confused " refers to the act of knowledge or to the way in which the intellect grasps its object. As such, is must be distinguished from a " confused object." Of course where the object itself is confused, the act of apprehension can not be said to be perfectly clear either. While there is some interdependence, however, the two are not simply to be identified. Confer *Oxon.* 1, d. 3, q. 2, n. 22; IX, 48a: Aliud est confuse intelligere, et aliud confusum intelligere. Confusum enim idem est quod indistinctum et sicut est duplex indistinguibilitas ad propositum, scilicet totius essentialis in partes essentiales, et totius universalis in partes subjectivas, ita est duplex distinctio duplicis totius praedicti ad suas partes. Confusam igitur intelligitur, quando intelligitur aliquid indistinctum altero praedictorum modorum. Sed confuse dicitur aliquid concipi, quando concipitur sicut exprimitur per nomen; distincte vero quando concipitur, sicut exprimitur per definitionem.

[12] *Ibid.* n. 21; 48a: Confuse dicitur aliquid concipi, quando concipitur sicut exprimitur per nomen; distincte vero quando concipitur, sicut exprimitur per definitionem: *Ibid.* n. 24; 50a: Cognoscere distincte habetur per definitionem, quae inquiritur per viam divisionis, incipiendo ab ente usque ad conceptum definiti; *Ibid.* 49b: Nihil concipitur distincte, nisi quando concipiuntur omnia quae includuntur in ratione ejus essentiali.

signifies the thing immediately and not the concept.[13] A name therefore may stand for the entire thing and yet not tell us very much about it. Similarly, the confused concept may refer to the thing as a whole; it may include a wealth of intelligible data, but the content of the concept remains unanalyzed.

According to Scotus, then, what is first known by way of origin? His teaching can be summed up in three propositions. Absolutely speaking, our first knowledge will be of something confusedly known, since distinct knowledge is always preceded by some confused knowledge.[14] The first thing known confusedly will be the *species specialissima* whose singular, whether audible, visible or tangible, more efficaciously and strongly first moves the senses.[15] The first distinct knowledge will be that of "being".[16] To explain further:

In discussing this problem Scotus accepts provisionally that the intellect does not immediately grasp the singularity as such but only the common intelligible elements. The problem of an intuition of the singular seems to be in the back of his mind, but he does not wish to complicate the discussion by introducing it here.[17] If we say that singularity as such is not immediately

---

[13] For Scotus, as for Ockham, the concept and the spoken or written word all signify immediately the thing by a sort of parallel signification. Confer *Oxon.* 1, d. 27, q. 3, n. 19; X, 377b-378a: Licet magna altercatio fiat de voce, utrum sit signum rei vel conceptus, tamen breviter concedendo quod illud quod significatur per vocem proprie est res, sunt tamen signa multa ordinata ejusdem significati, *littera, vox* et *conceptus,* sicut sunt multi effectus ordinati ejusdem causae, quorum nullus est causa alterius. —See also *Repor. Par.* 1, d. 27, q. 2, n. 8; XXII, 334b: Isto modo signum se habet ad signatum, nam passio in anima, et verbum vocale, immediate sunt signa rei, unum tamen immediatius, puta passio, vox est signum remotius, et res tantum est, quae immediate significatur per utrumque.

[14] *Oxon.* 1, d. 3, q. 2, n. 27; IX, 51a: Confuse cognoscere est ante quodcumque distincte intelligere.

[15] *Ibid.,* n. 22; 48a: Dico, quod primum actualiter cognitum confuse est species specialissima, cujus singulare efficacius et fortius primo movet sensum, sive sit audibile, sive visibile, sive tangibile.

[16] *Ibid.* n. 24; 50a: Ens est primus conceptus distincte conceptibilis.

[17] *Ibid.* n. 22; 48a: Et hoc supposito quod singulare non possit intelligi sub propria ratione, de quo alias. Loquor enim modo de illis, quae certum

intelligible, our first confused knowledge could be said to include the total intelligible content of whatever accidental quality most forcibly affects our senses. This seems to be what is meant by the *species specialissima* or the *infima species*. We are hardly justified in referring it to the extra-mental object as a whole, namely, to the composite of substance and its accidents—for instance, a man, or a bird—as Belmond seems to do.[18] Certainly the note of substance or substantiality would have to be included in the *infima species* of man or bird. Yet Scotus never tires of pointing out that we have no immediate knowledge of substance as such.[19] Even the substantial character of the soul

est posse intelligi secundum omnem opinionem. Confer *Metaphy.* 7, q. 15, VII, 434-440.

[18] Séraphin Belmond, "Essai de Synthèse philosophique du Scotisme" in *La France Franciscaine,* XVI (1933), 97: ...l'entendement proprement dit entre directement en contact avec ce qui est essentiel à l'espèce; le *quod quid est,* ce par quoi un être est spécifiquement tel: homme, oiseau, cheval, etc.... C'est ce que Scot nomme 'image très spéciale.' Timotheus Barth seems to come closer to the truth. Confer "Die Stellung der univocatio im Verlauf der Gotteserkenntnis nach der Lehre des Duns Skotus" in *Wissenschaft und Weisheit,* V (1938), 244 (footnote): Dass die species specialissima der quidditas rei sensibilis materialiter gleichkommt, ergibt sich daraus, dass jene auch aus dem mundus materialis genommen wird. Die species specialissima ist nämlich das Resultat eines singulären Hör-, Seh- oder Tastvorgangs, der von der Aussenwelt, auf uns einwirkt. Weil der Inhalt, dieses Vorgangs nicht in seinem singulärem An-sich-sein erfasst werden kann, sondern nur nach seinen allgemeinen Umrissen, nämlich nach denen der species specialissima, so erstreckt sich unsere erste unmittelbare Erkenntnis der Dinge auf etwas, was mit der quidditas oder species der Dinge zusammenhängt.

[19] *Metaphys.* 7, q. 3, n. 3; VII, 362: Substantia de se est perfectius cognoscibile et in se et a nobis, si possemus ad illam pertingere, sed non possumus in vita ista. *Ibid.* n. 2, 361b: Species accidentium sunt in virtute phantastica, et illorum est abstractio, quae sunt ibi. Quod autem species substantiae non sit in intellectu probatio; quia tunc species substantiae prius esset in sensu, et sic posset substantia cognosci a sensu, cujus oppositum manifestum est...*Ibid.* n. 3: Dicitur, quod non est in intellectu nisi species accidentis, et immediate causatur illa species ab accidente et mediate a substantia, et primo repraesentat accidens, et secundario substantiam. Sed contra: species non potest repraesentare distincte illud, quod est perfectius illo, cujus est species. *Ibid.* 2, q. 3, n. 19; 110ab: In

eludes our immediate intuition in this life.[20] Neither in a physical nor in a metaphysical sense, neither from a philosophical nor from a scientific standpoint do we get the *species specialissima* of the physical substance in this first confused knowledge. When a physicist feels the mist on his cheek on a foggy morning, he does not grasp even confusedly that this vapor is composed of minute molecules, each of which is a triatomic polar compound whose hydrogen components are probably resonating between an ionic and a covalent linkage. If the philosopher can speak of the *species specialissima* of man in the sense of a rational, sensitive, living, corporeal substance, it is only because a great deal of reasoning and innumerable abstractions from phantasms have preceded it.

It is interesting to note that Scotus emphasizes that the attention of the intellect is called, as it were, to that element which makes the greatest impression on the senses. For instance, if the microscope is not sharply focused on a drop from a culture of protozoa, the first impression the observer will probably get is of movement. The other characteristics of the blurred, almost colorless shapes flitting over the field hardly register. If the observer were to look up at this moment, he would have only a confused impression of motion. Yet this is sufficient to form a distinct concept, which follows almost instantaneously upon the first concept. If asked, " What did you see? " he would answer, " I saw something." If asked further, he might specify it as a movement. He might infer, almost instantaneously from force of habit, that movement implies something in motion, and consequently, that he saw a moving object. The thing to note here is that at least the first element of distinct knowledge —

---

intellectu notitia visionis vel intuitiva, quae est prima cognitio, non est in via possibilis.... Secunda cognitio [sc. per speciem] est accidentium sensibilium solum, quia illa sola faciunt speciem in intellectu, non autem substantiae ut patet supra. In the *Opus Oxoniense* (lib. 1, d. 22, q. 1, n. 2; X, 224) Scotus shows how we arrive at the notion of substance by discursive reasoning.

[20] *Oxon.* prolog. q. 1, n. 11; VIII, 21b: Non enim cognoscitur anima a nobis, nec natura nostra pro statu isto, nisi sub ratione aliqua generali abstrahibili a sensibilibus.

the concept of "being" or "something"—does not necessarily presuppose an elaborate intelligible content in the first confused concept.

Whether the content of the phantasm is rich or meager, the mind tends to grasp confusedly the whole intelligible content or the *species specialissima*. This is followed sooner or later by the first distinct notion, the notion of "being" or "something."

THE CONCEPT OF BEING

This notion of being is not vague or indistinct, for the simple reason that it contains but one simple intelligible note or *ratio*.[21] Consequently, either I know it or I am in total ignorance of it.[22] There is no half-way stage. This "being" is *quasi-notissimum*, as St. Thomas puts it.[23] It can neither be properly defined,[24] nor can it be explained in the strict sense, since there is really nothing more readily known which could serve to elucidate it.[25] It can be conceived distinctly apart from any other concept, but all other distinct concepts presuppose it.[26]

The psychological accuracy of Scotus' analysis does not concern us here,[27] but we are interested in the light it sheds upon

---

21 *Oxon.* 1, d. 3, q. 2, n. 24; IX, 49b: Ens autem non potest concipi nisi distincte, quia habet conceptum simpliciter simplicem.

22 *Ibid.* q. 3, n. 12; IX, 111a: Tale simpliciter simplex, ignotum est omnino, nisi secundum se totum concipiatur.

23 *De Veritate,* q. 1, a. 1: Illud autem quod primo intellectus concipit quasi notissimum, et in quo omnes conceptiones resolvit, est ens.

24 *Oxon.* 1, d. 39, q. un., n. 13; X, 625a: Ens habet conceptum simpliciter simplicem, et ideo non potest esse [ejus] definitio.

25 *Ibid.* d. 2, q. 2, n. 31; VIII, 478b: Ens autem per nihil notius explicatur. *De Primo Principio,* c. 3, concl. 9 (ed. M. Mueller) 106.

26 *Oxon.* 1, d. 3, q. 2, n. 24; IX, 49b-50a: Ens autem non potest concipi nisi distincte, quia habet conceptum simpliciter simplicem; ergo potest ens est primus conceptus distincte conceptibilis.

27 For a fuller discussion of the problem and the arguments used by Scotus to substantiate his position see Basil Heiser, "The *Primum Cognitum* according to Duns Scotus" in *Franciscan Studies,* XXIII (1942), concipi distincte sine aliis, et alii non sine eo distincte concepto; ergo pp. 193-216.

the notion of transcendental being. For this notion is the most common or universal of all real notions.[28] It is obviously a real notion, since it is abstracted from a real entity and is predicable of what actually exists or at least can exist. There can be little doubt that this is the so-called "being" of metaphysics, for the second argument Scotus adduces to prove that being is the first distinct concept is based upon this assumption.[29]

In view of the recent trend of thought, developed principally by Maritain and to a lesser extent by Gilson,[30] the question arises, Is this notion of transcendental being to be considered primarily existential or essential? Since this transcendental notion of being of Scotus' is, to all appearances, to be identified with the being of metaphysics, the answer to this question will determine whether metaphysics is to be considered as existential or essential, in the sense coined by Maritain. St. Thomas and Aristotle are cited as exponents of an existential metaphysics; Scotus and Plato as advocating predominantly an essential metaphysics.[31]

In discussing this question, one important thing should be kept in mind. The problem of an essential or existential metaphysics is primarily a problem for Thomists or, more universally, for a system of philosophy which admits a real distinction between essence and existence—"a fiction," says Scotus, "of which I know nothing!"[32] Maritain unfortunately seems to have over-

---

[28] *Oxon.* 1, d. 3, q. 2, n. 24; IX, 49b: Primum sic conceptum est communissimum.

[29] *Ibid.* n. 25; 50a: Secundo probo idem, quia Metaphysica secundum Avicennam, ubi prius est prima secundum ordinem sciendi distincte, quia ipsa habet certificare principia aliarum scientiarum; ergo ejus cognoscibilia sunt prima distincte cognoscibilia. *Metaphy.* 2, q. 3, n. 21; VII, 111b.

[30] Jacques Maritain, *A Preface to Metaphysics* (New York, Sheed and Ward, 1943) second lecture, esp. pp. 19-25; Étienne Gilson, *Réalisme thomiste et critique de la connaissance* (Paris, J. Vrin, 1939), pp. 220-222; *God and Philosophy* (New Haven, Yale University Press, 1941), pp. 38-73.

[31] Maritain, *op. cit.*, pp. 20, 38; Gilson, *God and Philosophy*, pp. 63-69.

[32] *Oxon.* 4, d. 11, q. 3, n. 46; XVIII, 429a: Nescio enim istam fictionem, quod *esse* est quid superveniens essentiae.

looked this point in describing the "error which may be termed Platonic or Scotist." [33] As a result, he has given us a very ingenious delineation of what Scotus might have held had he been a Thomist.

What Scotus has actually done has been to give us an essential being that has lost none of its existential import. Since the position of Scotus on this matter has been treated already by Barth, there is no need to go into detail here.[34] We believe that Barth is essentially correct when he states that being, according to Scotus, represents primarily a quidditative notion but with a tendency or aptitude to exist.[35] Over and above the reasons listed by Barth for the quidditative nature of being, we call attention to the fact that being pertains to the order of distinct knowledge, namely, that kind of knowledge which is expressed by the definition.[36] Now the definition expresses the essential or quidditative elements of the thing, and being, as Scotus continually asserts, is the basic element in every essence and every definition.[37]

This "primacy of essence," Gilson suggests, "appears in the doctrine of Duns Scotus as a remnant of the Platonism anterior to Thomas Aquinas." [38] The real reason why Scotus maintains

---

[33] Maritain, *op. cit.*, p. 38.

[34] T. Barth, "De fundamento univocationis apud Joannem Duns Scotum" in *Antonianum,* XIV (July 1939), 277-287.

[35] *Ibid.*, 286f: Ideo ens quod in his quoque quaestionibus est fundamentum cognitionis Dei, comprehendit et notionem *quid* et *aptitudinem ad existentiam*...Et ista tendentia ad existentiam quae essentiae purae adiungitur, non destruit simplicitatem absolutam entis. Ratio est haec... inter essentiam et existentiam *aptitudinalem* non plus statuere potest quam distinctionem rationis *inadaequatam*."

[36] Confer note 12.

[37] *Oxon.* 1, d. 36, q. un., n. 11; X, 578b: Definitio est distincta cognitio definiti secundum omnes ejus partes essentiales. *Ibid.* d. 3, q. 2, n. 24; IX, 49b: Nihil concipitur distincte nisi quando concipiuntur omnia quae includuntur in ratione ejus essentiali; ens includitur quidditative in omnibus conceptibus quidditativis inferioribus.

[38] *God and Philosophy,* p. 69, footnote 17.

that the being of metaphysics is a quidditative notion, however, is to be sought not in Platonism, but in the simple Aristotelian axiom that no science of the contingent *qua* contingent is possible.[39] Since all existence, with the exception of God's existence, is contingent, metaphysics as a science of " existences " is unintelligible. Existence is as little capable of serving as the " stuff of which the universe is made " as the *élan vital* of Bergson or the eternal flux of Heraclitus. Maritain recognizes this difficulty when he insists, like Scotus, that we must abstract from actual existence.[40] To have a science, it is necessary to discover a necessary element in the contingent. The notion of actual existence (as we experience it) does not contain any such necessary element, but the notion of possible existence does contain an element of necessity. What actually exists (God alone excepted) is mutable, contingent and temporal; what can be is necessary, immutable and eternal. For this reason medieval physics could never be a science of motion, but a science of the *ens mobile*, namely, the immediate subject of motion. Similarly, metaphysics is not a science of " being " in the adverbial sense of exist-

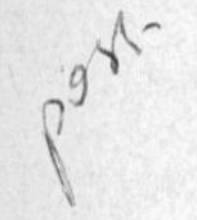

[39] Aristotle, *Analytica Posteriora* I, c. 6, 8 & 30 (passim); Scotus, *Oxon.* prol. q. 4, n. 41; VIII, 282b-283a: Scientia est necessariorum; ergo circa contingentia non est scientia. Antecedens patet ex definitione *scire* 1 *Poster.* similiter 7 *Ethic.* ...1 Posterior. "Scientia est necessarii dicti de contingente."...; *Reportata Par.* prol., q. 1, a. 1, n. 4; XXII, 8a: Secunda conditio [scientiae], scilicet quod sit *veri necessarii,* sequitur ex prima, quia si scientia esset veri contingentis, posset sibi subesse falsum, propter mutationem objecti, sicut opinioni.... Et hoc est quod dicit Philosophus 7 *Metaphy.* text. 53 et 1 *Posterior.* text. 21: "Corruptibilium non est demonstratio," quia sicut non contingit scientiam quandoque esse ignotam, ita nec demonstrationem quandoque non esse demonstrationem; contingeret autem utrumque, si esset non necessarium, sed contingens.—Confer also *Quodl.* q. 7, n. 9; XXV, 290b.

[40] *Metaph.* 7, q. 2, n. 6; VII, 328a: Scientia autem stricte sumpta per demonstrationem simpliciter abstrahit ab actuali existentia.—Maritain, *Preface to Metaphysics,* pp. 21-22: Therefore where existence is contingent, simply posited as a fact, as is the case with all created being, it must, because of this defect in its object, be directly orientated only to possible existence...It considers the essences as capable of actualisation, of being posited outside the mind.

ing, but in the nominative sense of " a being " or the immediate subject of existence, that is, " the existible."

It is this idea that Scotus seeks to bring out when he " defines " being as " that to which to be is not repugnant ".[41] To call this quidditative notion a pure essence, in the sense of Maritain, and to treat it as a sort of " least common denominator " between the real and the logical order, is an inexcusable perversion of the conception Scotus had of being. The term " to be " (*esse*) is to be understood in the sense of actual existence. Whenever it is to be understood of any other kind of existence, for instance, mental existence, Scotus carefully qualifies the term.[42] He speaks, for instance, of the *esse diminutum, esse cognitum,* etc. He also recognized that the terms " being," "quiddity," "thing," etc. are used equivocally and can be applied both to real and logical entities. But he carefully distinguishes between the two orders.[43] Only where the note of compatibility with real existence is to be found do we have a notion of real being or real thing.[44] And metaphysics differs from logic precisely in this, that the former is a real science and deals with real being; logic, on the contrary, deals with logical or mental entities.[45]

It is also clear that if being is the first element of distinct knowledge, it must naturally be one which includes the subject of existence rather than actual existence. For we cannot conceive existence save in reference to a subject. But when we

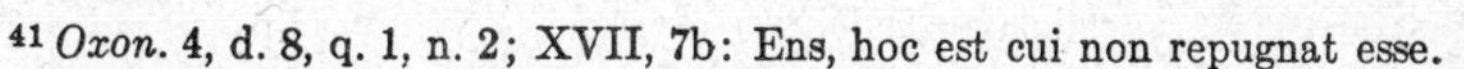

[41] *Oxon.* 4, d. 8, q. 1, n. 2; XVII, 7b: Ens, hoc est cui non repugnat esse.

[42] Confer for instance *Oxon.* 1, d. 36, q. un. n. 6; X, 575a.

[43] *Quodl.* q. 3, n. 2; XXV, 113b-114b.—Confer Ch. I, note 13, p. 4.

[44] *Ibid.* Accipitur [i. e. ens vel res] pro ente, quod habet vel habere potest aliquam entitatem non ex consideratione intellectus.

[45] *Metaphy.* 6, q. 3, n. 15; VII, 346a: Ens diminutum. . . est ens Logicum proprie. Unde omnes intentiones secundae de tali ente praedicantur, et ideo proprie excluditur a Metaphysico. Convertitur [hoc ens diminutum] tamen cum ente aliqualiter, quia Logicus considerat omnia, ut Metaphysicus. Sed modus alius considerationis, scilicet [Metaphysicus] per quid reale, et [Logicus] per intentionem secundam, sicut convertibilitas entis simpliciter et diminuti, quia neutrum alterum excedit in communitate. Quidquid enim est simpliciter ens, potest esse ens diminutum.

attempt to determine the precise whatness or quiddity of this subject, we include a reference to actual existence. It is that which is compatible with actual existence. Our first answer to the question, What is it? (*Quid est?*), is that the object first perceived confusedly is definitely designated as a thing or a being, namely, a subject capable of being.

Scotus, it should be noted, has used the negative formulation: *ens, hoc est cui non repugnat esse.* This is to call attention to the fact that the "possibility of existence" is to be understood in the sense of logical possibility, that is, a compatibility of notes.[46] This definition, consequently, is verified in that which actually exists (actual being) or also in that which, while it does not actually exist, can exist.

To sum up, "being" is primarily a quidditative concept. It is the first answer to the question: *Quid est?* It is an irreducibly simple concept (*simpliciter simplex*) in the sense that it cannot be broken down into two more simple concepts.[47] It expresses in a simple indivisible notion what we express through the phrase "subject capable of existence."[48] It is not a full concept or a rich concept. On the contrary it has the barest content, a minimum in the order of comprehension. In the language of the logician, it expresses any subject of which existence may be predicated in the mode of possibility. In no sense is this *concept* of being a "mystery" to be explored. In the language of St. Thomas, it is *quasi-notissimum.* Neither is it like the "empty" hat of the magician which by some metaphysical legerdemain can be made to produce all other concepts. Hence it is to be distinguished from the "confused knowledge" that precedes. The

---

[46] *Oxon.* 1, d. 2, q. 7, n. 10; VIII, 529b.

[47] *Oxon.* 1, d. 3, q. 2, n. 21; IX, 47b–48a: Alius est conceptus simpliciter simplex, et alius est conceptus simplex, qui non est simpliciter simplex. Conceptum simpliciter simplicem voco, qui non est resolubilis in plures conceptus, ut conceptus entis et ultimae differentiae. Conceptus simplex, sed non tamen simpliciter simplex, est quicumque potest concipi ab intellectu actu simplicis intelligentiae, licet possit resolvi in plures conceptus seorsum conceptibiles, sicut est conceptus definiti, vel speciei. *Oxon.* 1, d. 2, q. 2, n. 5; VIII, 406b; *Coll.* 35, n. 7, V, 290bff.

[48] Confer note 41.

latter is much fuller in content. It contains both common and differential elements but in an unanalyzed state. It can be explored by definitive notions, and the first step in the direction of such exploration is the predication of the concept of being. While the confused concept contains the intelligible elements revealed immediately in the phantasm through the activity of the agent intellect, it does not contain the notion of substance *as such* any more than it contains a notion proper to God. Nevertheless, it does contain the necessary data with which to reason to the existence of substance, since this first distinct concept of being is univocally common to substance and accident. Being thus becomes the core around which the mind is able to build up a proper notion of the whole object whose sensible qualities are immediately intelligible in the phantasm.[49]

In understanding Scotus, then, it is important to distinguish between the distinct notion of being and its formal content, the confused concept which precedes, and finally the physical thing or *res* to which both concepts refer.

THE PRIMARY OR ADEQUATE OBJECT OF THE INTELLECT

The adequate object of a cognitive power is known as its first, or better, its primary object. Adequate means properly proportioned or commensurate to the power in question.[50] It implies that the object is able to motivate or elicit from the faculty an awareness of itself both as to its formal content and its virtual implications.[51] It must be either formally or virtually

---

[49] Confer preceding chapter, p. 52-53.

[50] *Oxon.* 2, d. 3, q. 9, n. 12; XII, 216.

[51] *Oxon.* 1, d. 3, q. 2, n. 21; IX, 47b: Tertius est ordo adaequationis vel causalitatis praecise...de tertia autem primitate habetur 1 *Posteriorum* in definitione universalis, quia primo dicit ibi praecisionem et adaequationem. *Ibid.* q. 3, n. 3; 89b-90a: Nulla potentia potest cognoscere objectum aliquod sub ratione communiori, quam sit ratio sui primi objecti, quod patet primo per rationem, quia tunc illa ratio primi objecti non esset adaequata.... Quidquid per se cognoscitur a potentia cognitiva, vel est ejus objectum primum, vel continetur sub illo objecto. *Ibid.* n. 5; 98a: Primum objectum naturale alicujus potentiae habet naturalem ordinem ad illam potentiam...sub ratione motivi. *Oxon.* prol. q. 3, n. 5; VIII, 123b: Proportio objecti ad potentiam est proportio motivi ad mobile, vel activi ad passivum. *Ibid.* nn. 22-23; 174-175.

coextensive with whatever can be known by the particular power under consideration, so that by knowing it fully everything is known that can be known.[52]

An object is said to be adequate or commensurate to a respective faculty in either of two ways: a) in the sense that it enjoys a *primacy of virtuality*[53] or b) a *primacy of commonness*[54] in regard to whatever can be known. An example of the first would be God's essence in reference to His divine intellect.[55] For His essence alone is capable of motivating His intellect. Yet He knows not only His own essence but He knows also the creatures, which are really distinct from Him. His essence has a primacy of virtuality, for it has the power or *virtus* of producing, as it were, a knowledge of all that can be known. An example of a primacy of commonness would be color in reference to the faculty of sight. In the primacy of commonness, the " object " is not a particular physical entity but rather a common concept or *ratio* which can be predicated of all the physical objects capable of motivating the faculty.[56]

In the third question of this distinction[57] Scotus attempts to determine the adequate object of the human intellect. Because of a great deal of misunderstanding regarding Scotus' position, it will be well to preface a remark to the bare summary of his

---

[52] *Oxon.* 1, d. 3, q. 3, n. 5; IX, 98a.

[53] *Ibid.*: . . habet primitatem adaequationis . . . propter virtualitatem, quia scilicet virtualiter continet omnia in se per se intelligibilia. . . . ita quod . . . . movet intellectum nostrum primo ad se, secundo ad omnia alia cognoscibilia cognoscenda. *Oxon.* prol. q. 3, n. 22, VIII, 174b.

[54] *Ibid.* d. 3, q. 3, n. 5; IX, 98a: . . habet primitatem adaequationis propter communitatem, ita quod dicatur de omni objecto per se intelligibili a nobis.

[55] *Ibid.* 98ab: Sicut autem praedictum est in quaestione de subjecto Theologiae, essentia divina ideo est primum objectum intellectus divini, quia ipsa sola movet intellectum divinum et ad cognoscendum se, et omnia alia cognoscibilia ab ipso intellectu.

[56] *Rep. Par.* prol. q. 1, a. 2, n. 7; XXII, 10ab: Non potest esse aliquod unum objectum adaequatum illi potentiae et habitui, sed aliquod commune, cujus ratio salvatur in omnibus illis, quae per se movent vel terminant.

[57] *Oxon.* 1, d. 3, q. 3; IX, 87ff.

teaching presented here.[58] The problem of the adequate object of the intellect must not be confused with the problem of being as the subject of metaphysics, as has been done, for example, by Gilson and others.[59] The solution of the latter has nothing to do with revelation, whereas a complete solution of the former is possible only by taking into account what we know on faith.[60]

Scotus criticizes two current opinions. The first could be called Aristotelian; it makes the quiddity of a material or sensible being the proper object of the human intellect.[61] The second is a form of Augustinian illuminationism championed by Henry of Ghent, which makes God the proper and adequate object of the intellect. While Scotus is primarily concerned with determining the adequate object of the intellect in its present state of existence,[62] still in discussing the Aristotelian position, he finds that revelation throws a great deal of light on certain anomalies regarding the intellect's present state of existence. How-

---

[58] For a more detailed account confer Basil Heiser, "The *Primum Cognitum* according to Duns Scotus" in *Franciscan Studies,* XXIII (1942), 193-216.

[59] Confer for instance, É. Gilson, "Les seize premiers Theoremata et la pensée de Duns Scot" in *Archives d'Histoire doctrinale et littéraire du Moyen Age,* XII-XIII (1937-1938), esp. pp. 82-3: Mais l'*Opus Oxoniense* lui-même nous apprend, que l'intellect humain ne serait même pas capable, naturellement et sans l'aide de la Révélation, de s'élever à la véritable notion d'être, qui est l'object de la metaphysique. Le demi-succès d'Avicenne lui-même, cet infidele, ne s'expliquerait pas si l'on ne savait qu'il s'est inspiré d'une théologie. Puis-qu'il en est ainsi, comment ne pas voir que toute la métaphysique de l'*Opus Oxoniense* repose sur la saisie d'un objet que, bien que ce soit son objet naturel, l'intellect de la nature déchue n'est plus capable, à lui seul, de découvrir?—B. Heiser seems to follow Gilson in this matter. Confer his "The Metaphysics of Duns Scotus" in *Franciscan Studies,* XXIII (1942) p. 349, footnote n. 72.

[60] We intend to deal with this subject more in detail in a forthcoming article in the *Franciscan Studies.*

[61] *De Anima,* III, 7 (431a 14-17) Actually, Aristotle never seems to answer the question he raises in 431b 18.

[62] *Oxon.* 1, d. 3, q. 3; IX, 87ab: Juxta illud quod tactum est in tertio articulo secundae questionis, de objecto adaequato intellectus et praeciso, quaeritur, utrum Deus sit objectum naturale primum, hoc est, adaequatum respectu intellectus viatoris?

ever, the part that revelation plays in the ultimate solution of the problem should not blind us to what can be proved by natural reason alone. For the notion of adequate object contains at least two distinct requirements. First, the adequate object must be able to motivate the intellect. Secondly, it must be formally or virtually coextensive with whatever the intellect is able to know. Reason alone cannot tell us a great deal about the former, but it can about the latter. And it is the latter that interests us here.

At first sight, Aristotle might seem to be correct, since, in this life at least, only material quiddities seem to have the power to motivate our intellect.[63] Yet it is quite clear that material quiddities cannot be the adequate object of the intellect in the same sense that color is the adequate object of vision. For while color can be predicated of whatever can be seen, sensible quiddity cannot be predicated of whatever can be known. We have a very real and a valid notion of God and substance. We are able to form a concept that is more universal than sensible quiddities and is predicable of spiritual beings as well. The very fact that we possess a science of metaphysics—a science which transcends the sensible order and deals with being qua being—is proof that sensible or material quiddities are not commensurate with our intellect. In short, we are in the anomalous position of possessing a faculty which is able to transcend its starting point and arrive at a notion of things whose proper *ratio* is neither formally nor virtually contained in the simple apprehension of the sensible quiddity. The eye never transcends color. It never sees anything that is more common and universal than its proper object.[64] Consequently, on the basis of reason alone, Scotus argues, we have

---

[63] *Oxon.* 1, d. 3, q. 3, n. 24; IX, 148b: Primum objectum adaequatum sibi in movendo pro statu isto sit quidditas rei sensibilis. *Oxon.* prol. q. 1, n. 12; VIII, 22b.

[64] *Ibid.* 1, d. 3, q. 3, n. 3; IX, 89b: Visus non cognoscit aliquid sub ratione communiori, quae sit ratio coloris vel lucis, nec imaginatio aliquid sub ratione communiori, quam sit ratio imaginabilis, quod est primum objectum ejus; sed intellectus cognoscit aliquid sub ratione communiori, quam sit ratio entis materialis, quia cognoscit aliquid sub ratione entis in communi, alioquin Metaphysica nulla esset scientia intellectui nostro.

reason to question the Aristotelian hypothesis and to look for a broader and more universal notion than that of sensible quiddity as the adequate object of the intellect.[65]

On the other hand, the " ontologism " of Henry of Ghent is inacceptable. To begin with, God is not an object that is naturally ordained to our intellect, *sub ratione motivi.*[66] Such an ordination, as Scotus indicates elsewhere, militates against God's absolute independence of creatures.[67] If we say that God is included in the general notion of being, and to the extent that being

---

[65] We can show from reason that sensible quiddity is not properly speaking the adequate object of our intellect, since it is possible to form a more universal and abstract notion, namely, being qua being. On the other hand, we have no assurance that, if we were immediately confronted by an immaterial substance, the latter would be able to motivate our intellect without the medium of phantasm. Here is a point that has been misunderstood by Gilson (confer note 59). Far from saying that we need revelation to arrive at a concept of being qua being or that faith is needed to arrive at the knowledge of the object of metaphysics, Scotus argues that because we have a concept of being qua being and a natural science of metaphysics we have confirmatory arguments based on pure reason for what we know conclusively by faith. This is brought out both in the *Oxoniense* (loc. cit.) and even more forcefully in the *Reportata Parisiensia* 1, d. 3, q. 1, n. 3; XXII, 93b: Si dicas quod hoc est creditum . . . arguitur per rationem naturalem sic: Nulla potentia cognitiva potest cognoscere aliquid sub ratione illimitatiori, quam sit ratio sui primi objecti, quia si sic, illud non esset objectum primum, et sibi adaequatum; sed intellectus viatoris potest cognoscere ens quod est illimitatius quidditate materiali, aliter non posset habere cognitionem Metaphysicalem, ergo, etc.

[66] *Oxon.* 1, d. 3, q. 3, n. 5, IX, 98a: Primum objectum naturale alicujus potentiae habet naturalem ordinem ad illam potentiam; Deus non habet naturalem ordinem ad intellectum nostrum sub ratione motivi, nisi forte sub ratione alicujus generalis attributi, sicut ponit illa opinio; ergo non est objectum primum nisi sub ratione illius attributi.

[67] Gilson has shown how Scotus safeguards the independence of God in his excellent work, *The Spirit of Medieval Philosophy* (New York, Scribner's Sons, 1940) ch. 13, esp. p. 256. A free act or decree of God is required that the divine essence may motivate the created intellect. Confer *Quodl.* q. 14, n. 16; XXVI, 54a: Nullum igitur intellectum creatum movet essentia ut essentia tanquam motivum per modum naturae, sed omnem intellectionem illius essentiae, quam non causat aliquod creatum, causat immediate voluntas divina.

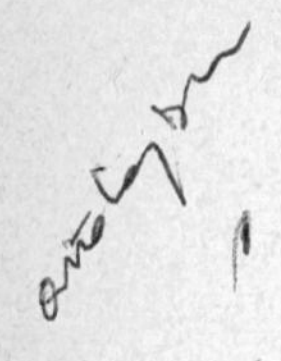

motivates our intellect, God may be said to be an object naturally ordained to it, then being rather than God should be called the adequate object.[68] Furthermore, God has a primacy neither of commonness nor of virtuality in regard to all those objects that of themselves are intelligible to our intellect. It is obvious that He cannot be predicated of all that we know, for that would be pantheism. If He possessed any primacy of adequation in regard to our intellect, it would have to be one of virtuality. But this is in patent contradiction to facts, for other things besides God are capable of moving our intellect *propria virtute*. In no sense can the divine essence be said to motivate us first to know itself and secondly all else besides God that can be known.[69]

For the same reasons, substance in general cannot be primary object of our intellect, since accidents in virtue of their own nature motivate the mind. Substance, therefore, does not primarily move the intellect to know it and through it to know all other things that are intelligible.[70]

---

[68] *Oxon.* 1, d. 3, q. 3, n. 5; IX, 98a: Sed particulare, quod non intelligitur nisi in aliquo communi, non est primum objectum intellectus, sed magis illud commune.

[69] *Ibid.* 98ab: Praeterea, certum est quod Deus non habet primitatem adaequationis propter communitatem, ita quod dicatur de omni objecto per se intelligibili a nobis; ergo si aliquam habet primitatem adaequationis, hoc erit propter virtualitatem, quia scilicet virtualiter continet omnia in se per se intelligibilia. Sed propter hoc non erit objectum primum adaequatum intellectui nostro, quia alia entia movent intellectum nostrum propria virtute, ita quod essentia divina non movet intellectum nostrum primo ad se, secundo ad omnia alia cognoscibilia cognoscenda; sicut autem praedictum est in quaestione de subjecto Theologiae, essentia divina ideo est primum objectum intellectus divini, quia ipsa sola movet intellectum divinum et ad cognoscendum se, et omnia alia cognoscibilia ad ipso intellectu.

[70] *Ibid.* 98b: Per easdem rationes probatur quod non potest poni primum objectum intellectus nostri, substantia in communi propter attributionem omnium accidentium ad substantiam, quia accidentia habent propriam virtutem motivam intellectus; ergo substantia non movet ad se et ad omnia alia cognoscibilia.

What, then, is the adequate object of our intellect? What can be said to be commensurate with everything that is *per se* intelligible to our intellect? There is no single object that is virtually coextensive with what we can know. Neither is there any formal notion that can be predicated commonly of all that we can know.[71] But by looking for something that combines a primacy of virtuality with that of common predication, it is possible to speak of some sort of adequate object in regard to the intellect.

It is in being that we find just such an object. For in being, Scotus points out, a twofold primacy concurs, one a primacy by way of common predication, the other a primacy of virtuality. Neither primacy taken singly is coextensive with all that can be known. Nevertheless, the combination of the two exhausts the realm of knowability.[72] Since the transcendental notions are all *per se* intelligible, an analysis of this twofold primacy of being will clarify the precise relation of the concept of being to the remaining transcendentals.

THE DOUBLE PRIMACY OF BEING

In his attack on the theory of Henry of Ghent, Scotus showed that a primacy of virtuality alone was out of the question in regard to our intellect, because there was no single object or kind of object that motivates our mind. On the contrary, there are many different kinds of objects that our intellect can know, and each object is knowable in virtue of itself and not in virtue of

---

[71] *Ibid.*, n. 6; 102a: Ad quaestionem igitur, respondeo quod nullum potest poni primum objectum intellectus nostri naturale, propter adaequationem talem virtualem, propter rationem tactam contra primitatem objecti virtualis in Deo vel in substantia. Vel igitur nullum ponetur primum objectum, vel oportet ponere primum adaequatum propter communitatem in ipso. *Ibid.* 102b: Dico quod ens non est univocum dictum in *quid* de omnibus per se intelligibilibus.

[72] *Ibid.* n. 12; 111b: In ente concurrat duplex primitas, scilicet primitas communitatis in *quid,* ad omnes conceptus non simpliciter simplices, et primitas virtualitatis in se vel in suis inferioribus, ad omnes conceptus in se simpliciter simplices. Et quod illa duplex primitas concurrens sufficiat ad hoc, quod ipsum sit primum objectum intellectus, licet neutram habeat praecise ad omnia per se intelligibilia.

something other than itself.[73] He proceeds to discuss why the concept of being is not predicable univocally of all that can be known and, consequently, why a primacy of commonness alone is insufficient. At the conclusion of this discussion we find the following résumé of his position.

> And I say that... since nothing can be more common than being, and that being cannot be predicated univocally and *in quid* of all that is of itself intelligible (because it cannot be predicated in this way of the ultimate differences or of its attributes), it follows that we have no object of the intellect that is primary by reason of a commonness *in quid* in regard to all that is of itself intelligible.
>
> And yet, notwithstanding, I say that being is the first object of the intellect, because in it a twofold primacy concurs, namely, a primacy of commonness and of virtuality. For whatever is of itself intelligible either includes essentially the notion of being or is contained virtually or essentially in something else which does include being essentially. For all genera, species, individuals, and the essential parts of genera, and the Uncreated Being all include being quiddita-tively. All the ultimate differences are included essentially in some of these. All the attributes of being are virtually included in being and in those things which come under being.
>
> Hence, all to which being is not univocal *in quid* are included in those to which being is univocal in this way. And so it is clear that being has a primacy of commonness in regard to the primary intelligibles, that is, to the quidditative concepts of the genera, species, individuals, and all their essential parts, and to the Uncreated Being. It has a virtual primacy in regard to the intelligible elements included in the first intelligibles, that is, in regard to the qualifying concepts of the ultimate differences and proper attributes....[74]

---

[73] *Ibid.* n. 5; 98a.

[74] *Ibid.* n. 8; 108b-109a; Dico quod ex istis quatuor rationibus sequitur, cum nihil possit esse communius ente, et ens non possit esse commune univocum dictum *in quid* de omnibus per se intelligibilibus, quia non de differentiis ultimis, nec de passionibus suis, sequitur quod nihil est primum objectum intellectus nostri propter communitatem ipsius *in quid* ad omne per se intelligibile; et tamen hoc non obstante, dico quod primum objectum intellectus nostri est ens, quia in ipso concurrit duplex primitas, scilicet

> And thus it appears how in *being* there concurs a tw[illegible] primacy, namely, a primacy of commonness *in quid* in r[illegible] to all concepts that are not irreducibly simple (*simpliciter simplices*) and a primacy of virtuality in itself or in its inferiors regarding all concepts which are irreducibly simple.[75]

To realize the full import of this passage, we must understand what Scotus means by such notions as *in quid* predication, ultimate differences, proper attributes, " virtually contains," and the like. Let us consider some of the more important ideas briefly.

1. *In quid predication*

In quid and *in quale* are two basic modes of predication. They refer primarily to the five predicables of Porphyry, namely, the genus, species, specific difference, property and accident, though Scotus extends the idea of an *in quid* and *in quale* predication to the transcendental order. Briefly, the difference between the two is this: to predicate *in quid* means to predicate either the entire essence (species) or at least the determinable part of the essence (genus). The term is derived from essence or *quiddity* and such predication represents an answer to the question: What is it? (*Quid est?*). To predicate *in quale* means to predicate a

---

communitatis, et virtualitatis. Nam omne per se intelligibile aut includit essentialiter rationem entis, vel continetur virtualiter, vel essentialiter in includente essentialiter rationem entis. Omnia enim genera et species, et individua, et omnes partes essentiales generum et ens increatum includunt ens quidditative. Omnes differentiae ultimae includuntur in aliquibus istorum essentialiter. Omnes passiones entis includuntur in ente et in suis inferioribus virtualiter. Igitur illa, quibus ens non est univocum [dictum] *in quid,* includuntur in illis quibus ens est sic univocum. Et ita patet, quod ens habet primitatem communitatis ad prima intelligibilia, hoc est, ad conceptus quidditativos generum et specierum et individuorum et partium essentialium omnium istorum et entis increati, et habet primitatem virtualitatis ad [omnia] intelligibilia inclusa in primis intelligibilibus, hoc est, ad conceptus qualitativos differentiarum ultimarum et passionum propriarum. [Note: This text as well as that of the following note have been corrected in the same manner as has the key-text. Confer chapter 1, note 14.]

[75] *Ibid.* n. 12, 111b: Ex his apparet quomodo in ente concurrat duplex primitas, videlicet primitas communitatis *in quid* ad omnes conceptus non simpliciter simplices, et primitas virtualitatis in se vel in suis inferioribus, ad omnes conceptus simpliciter simplices.

further determination or qualification of the essence. This qualification (*quale*) may be either essential (e. g. the specific difference) or non-essential (e. g. property or accident). Since the specific difference really is a part of the essence or quiddity, it is sometimes said to be predicated *in quale quid*, or *in quale substantiale*, in order to distinguish it from properties or accidents which are said to be predicated either *in quale accidentale* or simply *in quale*.[76]

To predicate something *in quid*, it is not sufficient that the predicate be an essential note, but that it be predicated *per modum essentiae, id est, per modum subsistentis*. Predication *in quale*, whether it be an essential qualification or not, is always predicated *per modum denominantis*.[77] Whatever is predicated *per modum subsistentis* is conceived as existing in itself and not merely inhering in another. From the viewpoint of grammar it will always be a *noun*. Thus " substance," " whiteness," " rationality," " rational animal," " life," if used as predicates in a proposition, would be said to be predicated *in quid*. They indicate the entire quiddity, or at least the determinable part of the essence of that subject of which they are predicated. Denominative terms are always derived terms, for instance, grammarian is derived from grammar; strong from strength.[78] All adjectives or modifiers are denominative terms; hence whatever is predicated *in quale*, that is, by way of qualification, can be said to be predicated *per modum denominantis*. Thus " substantial," " white," " rational," " living," when used as predicates in a proposition, would be predicated *in quale*.

---

[76] *Super universalia*, q. 28, nn. 2-3; I, 332ab; *ibid*. q. 12, nn. 5-7; 155b-156b, *Report. Par.* 1, d. 8, q. 5, n. 14; XXII, 170b-171a, etc.

[77] *Super. univer.* q. 12, n. 5-6; I, 155b: *Praedicari in quid* est praedicare essentiam subjecti, per modum essentiae, id est, per modum subsistentis, et non denominantis: et hoc contingit dupliciter; vel quod praedicet totam essentiam subjecti et sic est Species... Si vero partem essentiae, sic est Genus... *Praedicari in quale* est praedicari per modum denominantis, quod contingit dupliciter; vel quod praedicet subjecti essentiam, per modum denominantis; et tunc praedicatur *in quale substantiale* sive *essentiale*, et sic est Differentia... vel quod praedicet accidens per modum denominantis, et tunc praedicatur *in quale accidentale*...

[78] Aristotle, *De Categoriis*, c. 1 (1a 12-15).

Scotus, we said, extends the idea of *in quid* and *in quale* predication to the transcendental order.[79] Thus such qualifications as "one," "true," "infinite," "potential," etc. are predicated *in quale.* "Being," like any other term, can be predicated *per modum subsistentis* or *in quid,* or as we shall see later, it can be predicated *denominative.* Transcendental being, as we have seen in the preceding portion of this chapter, is taken as the common determinable element, the first and fundamental note of the essence of anything; it is the ultimate *subject* capable of existence, the ultimate *quid.* All these descriptions indicate that it is considered *per modum subsistentis,* that it is taken in a nominative sense, that it does not mean simply abstract *isness* but is understood in the concrete sense of *a being;* namely, a noun with a singular and a plural. In this sense, being is said to be predicated *in quid.* It does not express the whole essence or entity of that of which it is predicated; it only expresses the *ultimate determinable* and *common* element to be found in any thing or in any notion that is capable of being resolved into several more simple elements. This brings us to the next distinction.

2. The *simpliciter simplex* concept

Simple concepts are concepts which result from a simple act of apprehension. Not every simple concept, however, is *simpliciter simplex,* that is, irreducibly simple. A concept is said to be irreducibly simple if it is incapable of further analysis, that is to say, if it cannot be broken down into two simpler concepts, one of which is determinable, the other determining.[80]

A *simpliciter simplex,* or irreducibly simple, concept is opposed, on the one hand, to a composite concept and, on the other, to those simple concepts which contain a number of intelligible elements, each of which could be conceived distinctly but which de facto are grasped in a single simple act. Such a simple concept, for instance, is the confused concept of the *species specialissima* mentioned above.[81]

---

[79] *Oxon.* 1, d. 8, q. 3, n. 22; IX, 615a-616a.

[80] Confer note 47.

[81] Confer p. 62 ff.

Composite concepts result from a synthetic activity of the mind. The composite concepts of God and substance have already been mentioned.[82] But any concept that contains several distinct elements, one of which is common and determinable (a *quid*), the other differential and determining (a *quale*), is a composite concept.[83] Thus, for instance, "rational animal," "living substance," etc. are composite concepts. The analysis of the intelligible content of any confused concept, or of the being it represents, and the resynthesis of those notes in the form of a definition will always yield a composite concept.

Note that Scotus is interested here only in whether or not the concept in question is irreducibly simple. Being is contained *in quid* and univocally in every concept that is not irreducibly simple. It does not matter whether being be conceived distinctly, as it is, for example, in the composite concept, "finite being," or whether it be conceived only confusedly, as in the simple concept of "redness."

3. *Ultimate differences*

Concepts that are not irreducibly simple can be analyzed or reduced to at least two more simple concepts, for instance, man can be reduced to "animal" and "rational." "Animal" in turn can be analyzed further into "sentient" and "organism." This process, however, cannot go on *ad infinitum*. Otherwise, as St. Thomas tells us, nothing would be known.[84] Ultimately we arrive at intelligible elements that are incapable of further analysis, and hence are *simpliciter simplices*. Such concepts since they have nothing in common, are what Aristotle calls

---

[82] Confer preceding chapter, pp. 54-56.

[83] *Oxon.* 1, d. 3, q. 3, n. 6; IX, 103a: Sicut ens compositum in re, componitur ex actu et potentia in re, ita conceptus compositus per se unus componitur ex conceptu potentiali et actuali, sive ex conceptu determinabili et determinante.

[84] *De Veritate,* q. 1, a. 1, c.: Quod sicut in demonstrabilibus oportet fieri reductionem in aliqua principia per se intellectui nota, ita investigando quid est unumquodque, alias utrobique in infinitum iretur, et sic periret omnino scientia et cognitio rerum.

"primarily diverse."[85] One element in every concept capable of further analysis will be common; the other, differential. The ultimate common notion which is *simpliciter simplex* is being. But where there is but one common and determinable element that is irreducibly simple, there are as many such differential elements as there are different concepts.[86] The concept of being is common, whereas they are unique; being is potential and determinable, whereas they are actual and lack all capacity of being further determined.

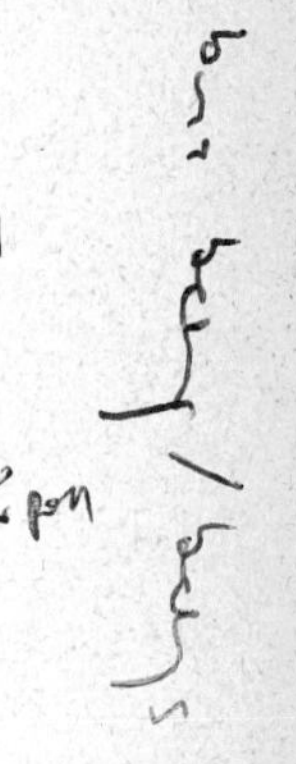

The meaning of the ultimate difference has been variously interpreted. Ockham, according to Boehner, conceives it as "the ultimate distinctive perfection in the quidditative order... It is that specific difference which cannot be determined further in the same order of quiddity, but which can be determined in another order, *viz.* of entity, by singularity."[87] Barth equates them with intrinsic modes, such as infinite and finite.[88]

Since Scotus gives as examples of concepts that are not irreducibly simple, that of the individual, the species, and the genus, and since every such concept must contain an ultimate difference, we may distinguish three types of ultimate differences: the *haecceitas* by which the species is ultimately differentiated, the specific differences which are irreducibly simple, and finally those transcendental differences which differentiate the various genera.

The *individuating differences* certainly are irreducibly simple. They are actual, where the concept of being is potential; they

---

[85] In the *Theoremata,* theor. XIII (V, 32-34) we have an excellent summary of the nature of the "*conceptus resolubilis*" and the "*conceptus simpliciter simplex.*" The doctrine agrees substantially with that of the definitely authentic works of Scotus.

[86] According to Aristotle two things are said to be *different* (τὰ διαφέροντα) only if they have something in common, if they have nothing in common they are simply *other* (ἕτερα). Confer *Metaphysica,* X, 3 (1054b 23-30). The ἕτερα were for the scholastics the "primo diversa."

[87] "Scotus' Teachings according to Ockham: I. On Univocity of Being" in *Franciscan Studies,* VI (1946), 102.

[88] Barth, "De fundamento univocationis apud J. Duns Scotum" in *Antonianum,* XIV (1939), 380ff.

are determining, where being is determinable, and so on. The only question which arises is whether or not Scotus has them in mind here. While they have a positive entity in themselves and are consequently *per se* intelligible absolutely speaking, nevertheless they are not intelligible as such in the present state of existence.[89] The solution of this question depends upon whether Scotus is speaking of the adequate object of the intellect precisely in so far as it is a faculty, (that is, independently of the present state of its existence), or whether he understands by adequate object, one that is commensurate with what is *per se* intelligible in our present state of existence. This latter problem, however, is too complicated to be discussed fully here, and is really of little moment so far as the purpose of this study goes. For if Scotus is referring to the *haecceitas* as an ultimate difference, the same relation will hold between the concept of being and the individuating differences as between being and the ultimate specific differences.[90]

Regarding the *specific differences*, it should be observed that not all are irreducibly simple and therefore all cannot be considered as ultimate differentiating elements. If we admit a plurality of forms, it will sometimes happen, says Scotus, that the reality or perfection from which the differential element is derived is not merely a distinct formality from that perfection to which the generic concept refers, but is a distinct physical being or *res*.[91] For instance, consider the concept " living body."

---

[89] Confer chapter 2, p. 29, note 51.

[90] *Oxon.* 2, d. 3, q. 6, n. 12; XII, 135a: Quoad hoc, ista realitas individui est similis realitati specificae.

[91] *Oxon.* 2, d. 3, q. 6, n. 12; XII, 134b: Dico quod illa realitas a qua sumitur differentia specifica est actualis respectu illius realitatis a qua sumitur genus vel ratio generis, ita quod haec realitas non est formaliter illa, alioquin in definitione esset nugatio, et solum genus sufficienter definiret, quia indicaret totam entitatem definiti sine differentia. Quandoque tamen illud contrahens est aliud a forma, a qua sumitur ratio generis, quando scilicet species addit rem aliquam super naturam generis, quandoque autem non est res alia, sed tantum alia formalitas, vel alius conceptus realis ejusdem rei; et secundum hoc aliqua differentia specifica habet conceptum non simpliciter simplicem, puta quae sumitur a forma; aliqua

" Living " is derived from a *form* really distinct from the body with its *forma corporeitatis*. Neither the body nor the life principle are entities that are primarily diverse. They are both beings, and hence " being " can be predicated *in quid* and univocally of both. Further analysis, however, will eventually yield a differential element that is *simpliciter simplex*.

Transcendental differences, such as infinite, necessary, etc. are among the primary determinations of being. They too fit the description of ultimate differences. It is true that Scotus refers to infinite-finite and necessary-contingent as intrinsic modes and denies that they are strict *differentiae*. By this, however, he does not wish to deny that such concepts are differential in character or that the reality signified by such concepts is *simpliciter simplex*. He merely calls attention to the great difference between the relationship of being and its primary determinations, compared to that which exists between the genus and specific difference.[92] Just because the elements of a composite concept like " infinite being " are related to one another as potency and act,[93] it does not follow that a corresponding metaphysical composition is to be found in God Himself.[94]

The entire class of disjunctive transcendentals would seem to fall directly under this classification of ultimate differences.

---

habet conceptum simpliciter simplicem, quae scilicet sumitur ab ultima abstractione formae; de qua distinctione differentiarum specierum dictum est *dist.* 3, *primi libri* [q. 3, n. 15] qualiter aliquae differentiae specificae includunt ens, et aliquae non.

[92] *Oxon.* 1, d. 8, q. 3, nn. 26-29; IX, 626-629.

[93] *Ibid.*, n. 28; 627b; *Collationes* in editione Waddingi non inclusae (ed. C. R. S. Harris, *Duns Scotus,* II) q. 3 pp. 372-373: Dicitur quod conceptus entis est determinabilis, sed non sic determinabile quod sit genus; quia tunc haberet limitationem determinatam, et potentialitatem extra rationem cujuslibet actus, et ipsum cum differentia sunt addita faciunt compositum. Nec est inconveniens, quod Deus sit compositus in descriptione quam habeo de eo in mente mea, sicut intelligendo eum ens primum et purum actum non possum resolvere conceptus istos in unum conceptum; et ideo in descriptione ejus habeo conceptum compositum, non tamen compositum sicut in definitione, sicut alii ponunt.

[94] *Ibid.*

However, in view of the fact that in disjunction each pair is coextensive with being, a case could be made for classifying them as proper attributes of being. Whether we prefer to consider the single members of the disjunction as ultimate differences of the respective being in which they are to be found, for instance, " infinite " in regard to God, or to consider the disjunction " infinite or finite " as an attribute of transcendental being itself, makes little difference. In either case, " being ", as a common univocal notion, is not predicable of either " infinite " or " finite " *in quid*.

The basic argument why " being " cannot be predicated univocally and *in quid* of the ultimate differences is that the concept of being is ultimate in the order of commonness, determinability and potentiality, whereas the final differences are ultimate in the order of uniqueness, determination and actuality. They are primarily diverse, not merely different,—to use Aristotle's distinction—and consequently have nothing in common so far as their proper perfection goes.[95] Being is only said to be common

[95] *Oxon.* 1, d. 3, q. 3, n. 6; IX, 102b-103a: Primum videlicet de differentiis ultimis, dupliciter probo. Primo sic: si differentiae includant ens univoce dictum de eis, et non sunt omnino idem; ergo sunt diversa, aliquid idem, entia; talia sunt proprie differentia ex 5 et 10 *Metaph.* ergo differentiae illae ultimae proprie erunt differentes; ergo aliis differentiis differunt. Quod illae aliae includunt ens quidditative, arguitur de eis, sicut de prioribus, et ita erit processus in infinitum in differentiis, vel stabitur ad aliquas omnino non includentes ens quidditative, quod est propositum, quia illae solae erunt ultimae. Secundo sic, sicut ens compositum in re componitur ex actu et potentia in re, ita conceptus compositus per se unus componitur ex conceptu potentiali et actuali, sive ex conceptu determinabili et determinante. Sicut igitur resolutio entium compositorum stat ultimo ad simpliciter simplicia scilicet ad actum ultimum et potentiam ultimam, quae sunt primo diversa, ita quod nihil unius includit aliquid alterius, alioquin hoc non esset primo actus, nec illud primo esset potentia, quod enim includit aliquid potentialitatis, non primo est actus: ita oportet in conceptibus, omnem conceptum non simpliciter simplicem, et tamen per se unum, resolvi in conceptum determinabilem et determinantem, ita quod ista resolutio stet ad conceptus simpliciter simplices, scilicet ad conceptum determinabilem tantum, ita quod nihil determinans includat, et ad conceptum determinantem, qui scilicet non includit aliquem conceptum determinabilem; ille conceptus tantum de-

or univocal in regard to infinite, finite, etc. in the sense that it is the common determinable to which both determining elements are ordained.[96]

Of these ultimate differences, the primary determinations (disjunctive transcendentals) obviously belong to the transcendental order. But also those specific differences which stand for pure perfections may be considered as transcendentals.

4. *Proper attributes*

Proper attributes or *propria* are those qualifications which are necessarily connected with their respective subject yet do not enter into its essential definition. That the angles of a triangle are equal to two right angles is a property of the Euclidean triangle. " Risibility " is a property of man. " Male " and " female " in disjunction are properties of animal. Similarly, " odd-even " in disjunction is a necessary attribute of number.[97] From the viewpoint of predication they are predicated necessarily (*per se*) but *secundo modo.* The notion of the subject is found in the definition of the predicate, whereas in predication *per se primo modo,* the predicate is either the whole or a part of the definition of the subject. Thus " male " cannot be defined without introducing the concept of animal, " risibility " without introducing the concept of man, and so on.

Being too has its proper attributes. Some like " one," " true " and so on, are coextensive with being; others, like " infinite-finite," "one-many," are proper only in disjunction. While it is possible to conceive the notion of being without conceiving the

terminabilis est conceptus entis, et determinans tantum est conceptus ultimae differentiae; ergo isti erunt primo diversi, ita quod unum nihil includit alterius.

[96] *Oxon.* 1, d. 3, q. 3, n. 12; IX, 111a: Est univocus eis, ut determinabilis ad determinantes, vel ut denominabilis ad denominantes. Unde breviter, ens est univocum omnibus, sed conceptibus non simpliciter simplicibus est univocum *in quid* dictum de eis, sed simpliciter simplicibus est univocus, ut determinabilis vel ut denominabilis, non autem ut dictum est de eis *in quid,* quia hoc includit contradictionem.—Note: the ultimate differences are said to *determine,* the attributes to *denominate* being.

[97] Aristotle, *Analytica Posteriora,* I, 4 (73b 10ff).

attribute, the attribute cannot be conceived without including the subject. Incidentally, this point has been overlooked in several recent works on Scotus, with the result that the whole concept of " virtual primacy " has been made unintelligible and even contradictory.[98] This non-mutual inclusion is evident in such notions as " one," " true," etc. They are secondary notions in the sense that they include over and above the mere notion of being some other *ratio*. " True," in the sense of intelligibility, is a relative notion. It presupposes two absolute notions, namely, the notions of being and of intellect. Even " one," which is considered an absolute attribute in contradistinction to " true " and " good," involves the relation of identity and diversity, as is clear from its definition as the " indivision of something in itself and its division from all else." [99]

Proper attributes, it was said, include the notion of the subject. This statement requires qualification. Number, for instance, does not enter into the definition of " odd " or " even " in the same way that " rational " or " animal " enter into the definition of man. If it did, the subject could be predicated of its attributes *per se primo modo*. On the contrary, "the subject enters into the definition of the attribute as it were by addition (ἐκ προσθέσεως). " [100] Scotus alludes here to the *Posterior Analytics* [101] and the *Metaphysics*,[102] where Aristotle discusses the nature of these attributes (τὰ συνδεδυασμένα) which pertain to their subject *per se* (καθ' αὑτό). We are unable to give a strict

---

[98] Harris, *Duns Scotus* (Oxford, Clarendon Press, 1927) II, pp. 66-7; M. Grajewski, *Formal Distinction of Duns Scotus* (Washington, Catholic University of America Press, 1944), p. 125; Barth, " De fundamento univocationis. . ." in *Antonianum*, XIV (1939), 378.

[99] *Oxon.* 4, d. 6, q. 1, n. 4; XVI, 532a: Unum est in se indivisum, et ab alio divisum.

[100] *Oxon.* 1, d. 3, q. 3, n. 6; IX, 103b: Passio per se secundo modo praedicatur de subjecto *primo Posteriorum;* ergo subjectum ponitur in definitione passionis sicut additum, ex eodem 1 *Post.* et ex 7 *Metaph.* Ens igitur in ratione suae passionis cadit ut additum.... ergo non est per se primo modo in ratione quidditativa earum.

[101] *Analytica Post.* I, c. 4.

[102] *Metaphysica,* VII, cc. 4-5.

definition of them, Aristotle explains, without mentioning the subject which they modify. Female, for example, is that quality by which an animal is female. But the peculiarity of such a quasi definition is not merely that it involves the introduction of something outside the term proper, namely, the subject which the term modifies, but that a tautology is involved. For we define X as it were by XY, and XY by XYZ, etc. "Consequently," Aristotle argues, "it is absurd that such things should have an essence; if they have, there will be an infinite regress."[103] "What I mean then 'by addition' (ἐκ προσθέσεως)," he continues, "is that it turns out that we are saying the same thing twice as in such expressions."[104] Such notions are essentially or radically qualifications (*qualia*). They have no common determinable element or *quid*. They also are *simpliciter simplices* in the sense that they cannot be analyzed into a distinct *quid* and a *quale*. For the common *quid* or subject which they modify, though introduced into their "definition", is not really a part of their proper entity. It is brought into it *by way of addition.*

Applying this to "being" and its "attributes," it must be said that "true," "good," "one," and the like are not properly "beings," nor can "being" be predicated of them *per se primo modo*. In scholastic terminology, they have no strict *quid* or essence, but are essentially qualifications irreducibly simple. They have only a *quid nominis*.

Shorn of logical terminology, the basic reason why the concept of being cannot be predicated univocally and *in quid* (that is, *per modum essentiae*) of its proper attributes or *quasi passiones* is because they have no strict essence or *quid* but are by their very nature a modification or mode of essence.[105]

---

[103] *Ibid.* (1030b 34): διὸ ἄτοπον τὸ ὑπάρχειν τοῖς τοιούτοις τὸ τί ἦν εἶναι· εἰ δὲ μή, εἰς ἄπειρον εἶσιν·

[104] *Ibid.* (1031a 4): τὸ δ' ἐκ προσθέσεως λέγω ἐν οἷς συμβαίνει δὶς τὸ αὐτὸ λέγειν, ὥσπερ ἐν τούτοις.

[105] *Oxon.* 1, d. 3, q. 3, nn. 6-7; IX, 103f.

5. *Virtual Primacy of Being*

The primacy of commonness, or better, of common predication, is clear enough. It simply means that wherever we have any notion that can be split up into a determinable and determining element, in the last analysis the determinable element will be the quidditative notion of being. Hence being is predicable of anything that can be grasped in a concept that is not *simpliciter simplex.*

The *primacy of virtuality,* however, has occasioned great difficulty. The fundamental idea behind "virtual containing" is that the object which is said to contain anything somehow possesses the *virtus* of producing the other.[106] God is said to contain created perfections, such as matter, composition, reasoning, etc., in the sense that He can produce them. This idea of production is carried over to the order of knowledge. Premises contain the conclusion virtually.[107] In some cases where the attributes can be deduced from their respective subject by using the definition of the latter as the middle term of the demonstration, the subject is said to contain the attribute virtually. For instance, the notion of a triangle can be used to demonstrate the attribute that its angles are equal to two right angles.[108] The subject of a science is said to contain all the truths of that science virtually, etc.[109]

There is no doubt that Scotus could have been much clearer on the meaning of virtuality. Ockham interpreted the virtual primacy of being to mean that from the *concept of being* the other notions could be deduced in some way. He criticized Scotus on this score.[110] Scotus, however, has never admitted that the attributes or differences could be extracted from concept of being, and in some places definitely denies that any deduction

---

[106] *Ibid.* q. 8, n. 1; IX, 398a: Continere autem virtualiter convenit causae activae.

[107] *Oxon.* prol, q. 3; n. 4; VIII, 122b-123a.

[108] Confer Aristotle, *De Anima,* I, 1 (402b 16ff).

[109] *Oxon.* prol. q. 3, n. 4; VIII, 122b-123a: Ratio primi subjecti [scientiae] est continere in se primo virtualiter omnes veritates illius habitus, cujus est.

[110] *Ordinatio,* prol. q. 5, art. 1, B ff.

of the sort is possible.[111] Furthermore, the extreme simplicity of the concept of being forbids any such interpretation. As we pointed out in the last chapter, a creature can cause a simple concept of being in the intellect, but it cannot cause a simple concept of "infinite." If the concept of being contained "infinite" virtually, could not a created object produce in our mind a simple notion of "infinite"?

Others, like Grajewski and Harris, have asserted that being contains virtually the ultimate differences and attributes in the sense that being cannot be conceived without these attributes or differences. But this interpretation is nowhere substantiated by texts and is in patent contradiction to the principle upon which the first proof of univocity is based.[112]

Barth has attempted to prove that "to contain virtually does not exclude, but on the contrary includes and presupposes 'to contain *in quid*'."[113] This would imply that being is either the whole or the determinable part of the essence of the perfections or attributes, or vice versa.[114] But this would mean that

---

[111] *Oxon.* 1, d. 39, n. 13; X, 625a: In passionibus autem disjunctis, licet illud totum disjunctum non possit demonstrari de ente.... Passiones autem entis convertibiles, ut communius immediate dicuntur de ente, quia ens habet conceptum simpliciter simplicem, et ideo non potest esse medium inter ipsum et suam passionem quia neutrius est definitio quae possit esse medium.

[112] Confer chapter 3, pp. 48-50.

[113] Barth, "De fundamento univocationis...," *Antonianum,* XIV (1939), 377: Virtualiter continere non excludit, immo includit et praesupponit τὸ continere in quid; τὸ continere virtualiter est prolongatio τοῦ continere in quid, addita ratione seu capacitate producendi. Si ergo ens suas passiones virtualiter continet, a fortiori eas in quid includere debet.

[114] Note, it is questionable whether to "contain something in *quid*" here means anything at all. When a predicate is predicated *in quid* of a given subject, it may be said *to be* contained *in quid.* That which contains something *in quid* has a greater, or at least equal, comprehension with that which it is said to contain. Man has a greater comprehension than animal. That which is contained may at most be synonymous with that which is said to contain it, but it can never be greater. Being, however, has only a single simple note in its concept. It is the least comprehensive (though it is correspondingly greatest in extension) of all notions. It *contains* only itself.

we have no other concept besides being that is *simpliciter simplex,* no concepts that are primarily diverse, and therefore no different concepts at all. It is equivalent to abandoning Scotus' position, as Barth himself admits.[115]

A much simpler and more intelligible interpretation is suggested by the context. The " being " that contains the attributes and differences virtually is not the formal concept or *ratio* of being at all, but that of which the concept of being is predicated. In other words, it is the *concrete object,* man, God, Peter, etc. or the *composite* concepts which represent such objects, namely, generic, specific, individual concepts, or even specific differences which are not irreducibly simple.[116]

Scotus tells us as much in the above passage, when he describes the double primacy of being. Being has a primacy of commonness regarding the " primary intelligibles." And what are the primary intelligibles? They are those objects which contain the concept of being *in quid* or essentially. They are, more specifically, those quidditative concepts of genera, species, individuals and their essential parts, together with the concept of the Uncreated Being. By essential parts Scotus could refer to physical parts, like matter and form, or even to those specific differences which are not *simpliciter simplices.* For he speaks of " quidditative concepts of the genera, species, etc." Those specific differences, however, which are *simpliciter simplices* have no strict quidditative concept but are simple qualifications (*qualia*).

And being, he goes on to say, " has a virtual primacy in regard to the intelligible elements included in the first intelligibles." Note that he does not say, being has a virtual primacy in regard to those things contained in the *concept of being,* but in the first intelligibles, namely, *in anything which includes the concept of being.* Or, as he said just before, being has a primacy of commonness and of virtuality because all *per se* intelligibles " either include essentially the notion of being "—and therefore fall under the primacy of commonness—" or are contained virtually or *essentially in something else which does include being essen-*

[115] *Ibid.* p. 391.

[116] Confer pages 78-79.

tially." The ultimate differences pertain to the essence and hence are " contained essentially," whereas attributes like " true," " one," etc., though necessarily connected with the essence, do not enter into its formal definition. Hence they are said to be " virtually contained . . . in something which contains being essentially." [117]

Further reasons for believing that it is the object of which being is predicated, rather than the concept of being itself which " virtually contains ", can be adduced. Scotus is attempting to determine the primary object which motivates our intellect. The content of a concept is an *ens rationis* and as such is incapable of motivating the intellect in the strict sense. Only an object is capable of motivating, or better, concurring causally in the formation of the concept, either in itself or through a likeness of itself (*species intelligibilis*).[118] To say that an object is adequate by reason of common predication, does not mean that the concept or *ratio as conceived* motivates the faculty, but that the physical entity of which it is predicable is able to motivate it. Color is the adequate object of sight by reason of common predication, but no eye ever saw " color " in the abstract, although it does see redness, blueness, et cetera—anything *in concreto* of which color can be formally predicated. Furthermore, the description Scotus gives of an object that is adequate by reason of virtuality does not apply to concepts but to physical entities.

---

[117] *Oxon.* 1, d. 3, q. 3, n. 5; IX, 108b-109a. While the concept of being does not contain the formal notions of the differential and attributive notions and has not of itself the *virtus* of producing them in the intellect, nevertheless, since the subject of a proposition that is *per se nota* is said to contain the predicate virtually if the latter is predicable in the second mode of per se predication (confer Scotus, *Metaphy.* 6, q. 1, 8; VII, 308b), even though the predicate does not enter into the formal definition of the subject, from the standpoint of predication being could be said to contain its proper attributes virtually. But this would not hold for the ultimate differences except where these could be reduced to a strict disjunction and predicated of being precisely as a disjunction.

[118] *Oxon.* 1, d. 3, q. 7, n. 20-21; IX, 361a-362b.

Though these ultimate differences and attributes lie outside the concept of being as the ultimate *quid* (for they are *simpliciter simplicia* and *primo diversa*), they do not lie outside the physical being or thing (*res*) but are unitively contained in it. Being, as a common quidditative notion, is only one of the many intelligible elements that can be abstracted from and predicated of the world of reality. As these thought elements are not simply formally identical, so neither is the precise reality which corresponds to each *a parte rei* absolutely identical. They are distinct in the sense that one formality may be said to be nonidentical with a second formality or with its intrinsic mode.[119]

The fact that Scotus draws a formal distinction, or at least a modal distinction, between being and its attributes,[120] gives rise to an objection. If being cannot be formally predicated *in quid* of the ultimate differences and proper attributes, then these qualifications lie outside the formal concept of being and therefore are non-being or formally "nothing." This objection was raised already by William of Alnwick,[121] Scotus' secretary; it reappears in the recent writings of Barth.[122] While this difficulty and its solution are not discussed in the *Opus Oxoniense*—though the basic principles for its solution are contained in the answer to some of the objections treated—the problem is explicitly raised elsewhere. In one of the *collationes* edited by Harris[123] the problem takes the form of the following objections.

---

119 Confer chapter 2, p. 30.

120 *Oxon.* 2, d. 16, q. un. n. 17; XIII, 43a: Ens continet multas passiones, quae non sunt res aliae ab ipso ente . . . distinguuntur tamen ab invicem formaliter et quidditative, et etiam ab ente. *Ibid.* 1, d. 8, q. 3, n. 27; IX, 627.

121 In a marginal note in the *Cod. Assis.* 172, f. 39v we find this note: "Ista opinio, licet sit vera quantum ad articulum primum [sc. quod ens sit univocum Deo et creaturae, cf. f. 39r], non tamen quantum ad secundum articulum [sc. quod ens non est univocum, dictum in quid de ultimis differentiis nec de propriis passionibus entis, cf. f. 39r]." (Quoted from T. Barth, "De univocatione etc." p. 379).

122 Barth, "De univocatione . . ." in *Antonianum,* XIV (1939) 391.

123 Harris, *Duns Scotus,* II, Appendix.

> If the common univocal concept is contracted, it is necessary that it be contracted by some addition. This addition is either a being or it is not a being. For it is necessary that whatever contracts another lie outside the notion [*ratio*] of that which it contracts. But nothing lies outside the notion of being. If it is not being, it does not contract.
>
> Likewise, if some concept were univocal, it would follow that nothing would be primarily diverse or distinct, because they would all agree in this common element and would be distinguished by other elements and these [other elements] would agree in being, and so ad infinitum.[124]

Note, the objection is used to disprove the assertion that being is a univocal concept. The same objection is raised by contemporary writers in the form of what we may call "the paradox of being."[125] Being is not only that by which things agree but it is also that by which they differ.[126] Consequently, being cannot be predicated in the same way of everything and therefore is not univocal but analogous.

Put in this form, it is quite apparent that the term 'being' must have several different meanings. Things cannot agree and differ by a formal ratio that is irreducibly simple. In the language of Aristotle, things differ only if they have something in common, that is, if the concept of each can be resolved into a

---

[124] *Collationes*, q. 3, (Harris, *op. cit.*, p. 373): Conceptus communis univocus si contrahatur, oportet quod contrahatur per aliquod additum, illud additum aut est ens, aut non ens; quia illud quod contrahit aliud, oportet quod sit extra rationem ejus. Sed nihil est extra rationem entis; si est non ens, non contrahit. Item, si sit aliquis conceptus univocus sequitur quod nulla erunt primo diversa vel distincta; quia semper conveniunt in illo communi, et distinguuntur per alia duo, et illa alia duo conveniunt in ente, et distinguuntur penes alia, et illa conveniunt in ente, et sic in infinitum.

[125] Confer, for instance, C. Bittle, *The Domain of Being* (Milwaukee, Bruce, 1942) p. 37-41.

[126] Bittle, *op. cit.* p. 39: "Hence, 'being' is not only that in which they agree; it is also that in which they *differ* and as such, then, it is not found in them in a strictly uniform manner."—Confer the excellent criticism of this "paradox" by J. J. Toohey, "The Term 'Being'" in *New Scholasticism*, XVI (1942) 107-129. If this objection to univocation held, it would be impossible to show that any predication is univocal.

common and a differential element. Otherwise they are simply diverse.[127] But being is by definition the common determinable and potential note. It is irreducibly simple. Hence it cannot be the formal differential element. Therefore, if being is predicable of the differential elements, that is, the ultimate differences and attributes, it is not this same quidditative being.

"Being", says Scotus, is predicable of the ultimate *qualia,* but not *in quid.* It is predicable *denominative.*[128] A denominative term, as has been said, is one derived from another term.[129] Boethius lists three conditions for a denominative term: 1) The derivative term must share in the same thing or *ratio* as that from which it is derived. "Just," for example, connotes somehow the *ratio* of justice. 2) It must participate in the same name. "Brawny" is named after "brawn," but "muscular" is not, although "brawny" and "muscular" connote the same basic *ratio.* 3) The derivative term must differ in case from the abstract or essential notion.[130]

---

[127] Confer note 85.

[128] *Collationes, ibid.:* Ad primum dicitur quod contrahitur, sed non per ens quidditative, loquendo de ultimis differentiis et primis, sed tantum per ens denominative dictum. Ad secundum, dico quod primo diversa sunt diversa, scilicet primo per differentias primas et ultimas; et de talibus non dicitur ens quidditative, sed denominative; et de aliis differentiis dicitur quidditative, et non conveniunt in ente quidditative, sed per accidens et denominative, quia tales differentiae, simplices et primae et ultimae, dicunt formaliter quale. Ens dicit quid, et ideo non dicitur in quid, nisi de eis quae formaliter dicunt quid, vel includunt quid· tales sunt differentiae interpositae inter primam et ultimam speciem, quae includunt duos conceptus, scilicet quid et quale.

[129] Aristotle, *De Categoriis,* c. 1.

[130] Boethius, *In Categorias Aristotelis,* lib. 1 (PL 64, 167-168): Denominativa vero dicuntur quaecumque ab aliquo solo differentia casu secundum nomen habent appellationem, ut a grammatica grammaticus, et a fortitudine fortis. Haec quoque definitio nihil habet obscurum. Casus enim antiqui nominabant aliquas nominum transfigurationes, ut a justitia justus, a fortitudine fortis, etc. Haec igitur nominis transfiguratio, casus ab antiquioribus vocabatur. Atque ideo quotiescunque aliqua res alia participiat, ipsa participatione sicut rem, ita quoque inde trahitur nomen dicitur enim justus. Ergo denominativa vocantur quaecunque a principali

All modifiers are denominative terms. It is interesting to note that such terms directly (*in recto*) designate the subject and connote *in obliquo* the formal *ratio* from which they are derived. For instance, " white " does not directly signify " whiteness " but the subject which possesses whiteness. This is evident from the proposition " That man is white." If whiteness were primarily designated by the predicate the proposition would be false, for " this man " is not " whiteness."

Applying this to being, it can be said that " of being," " pertaining to being," and the like are denominative forms of being. Similarly " being " in the sense of " existence " or " existing " is denominative. If it be objected that " ens denominatur ab esse ", it must be remembered that this is to be understood in a grammatical sense only. Existence and existing can neither be nor be conceived save as modes of some subject. The essential or quidditative notion is the original and nominative form.

Transcendental being, Scotus tells us, is defined as " that to which existence is not repugnant." This notion applies formally and properly to the primary subject of existence, and is expressed by the noun " a being." Qualifications and modifications are capable of existence only in so far as they inhere in a subject. Hence they are " a being " only in a derived sense, namely, in so far as they are unitively contained with the formal perfection of being in one and the same concrete physical entity. It is the physical *res* of which being is predicated that both agrees and differs from every other thing. But it does not differ because it possesses the formal perfection of being " a being," but because of other formalities or perfections whose formal *ratio* is to qualify beings and therefore are not " formally " *a being*, but are *of being*.

---

nomine solo casu, id est, sola transfiguratione discrepant. Nam cum sit nomen principale justitia, ab hoc transfiguratum nomen justus efficitur. Ergo illa sunt denominativa quaecunque a principali nomine solo casus id est sola nominis discrepantia, secundum principale nomen habent appellationem. Tria sunt autem necessaria, ut denominativa vocabulo constituantur: prius ut re participet, post ut nomine, postremo ut sit quaedam nominis transfiguratio, ut cum aliquid dicitur a fortitudine fortis, est enim quaedam fortitudo qua fortis participet, habet quoque nominis participationem fortis enim dicitur.

Hence it does not follow that whatever is not quidditatively being is nothing. Formal nothingness is neither quidditatively nor *denominative* a being.

This distinction between the *in quid* and *denominative* predication of being not only provides a very ingenious, yet basically simple, solution to the paradox of being. It is also typically characteristic of Scotus' philosophical attitude. In true Aristotelian fashion he is not content until he has subjected vague metaphysical notions to a rigorous logical analysis.

The relation of the transcendental notion of being to all other notions (both transcendental and predicamental) can be summed up best, perhaps, in diagrammatic form.

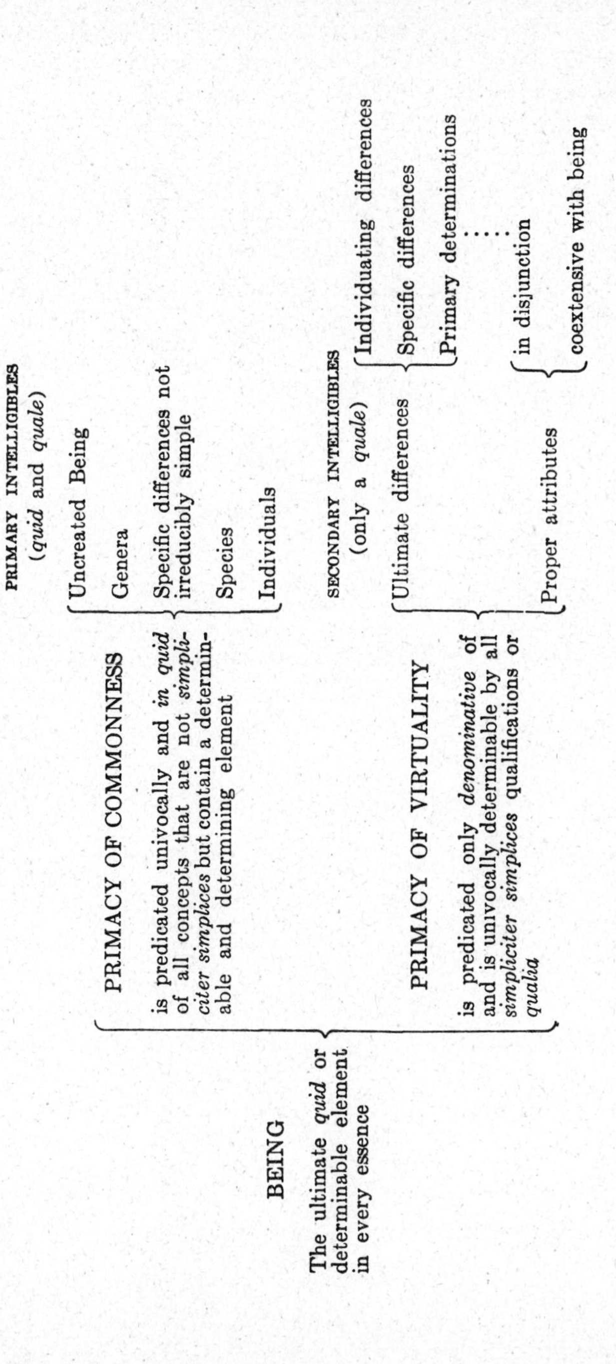

BEING AS THE ADEQUATE OBJECT OF THE INTELLECT
BEING
The ultimate *quid* or determinable element in every essence
PRIMACY OF COMMONNESS
is predicated univocally and *in quid* of all concepts that are not *simpliciter simplices* but contain a determinable and determining element
PRIMARY INTELLIGIBLES (*quid* and *quale*)
Uncreated Being
Genera
Specific differences not irreducibly simple
Species
Individuals
PRIMACY OF VIRTUALITY
is predicated only *denominative* of and is univocally determinable by all *simpliciter simplices* qualifications or *qualia*
SECONDARY INTELLIGIBLES (only a *quale*)
Ultimate differences
Individuating differences
Specific differences
Primary determinations
Proper attributes
in disjunction
coextensive with being

CHAPTER V

THE COEXTENSIVE TRANSCENDENTAL ATTRIBUTES

UNITY, truth and goodness are the transcendental attributes coextensive with being. Unlike Alexander of Hales, who identifies beauty with the *bonum honestum*,[1] Scotus does not attempt to add " the beautiful " as a distinct member of the trinity of coextensive transcendentals. Neither does he differentiate *aliquid*, *res* and *ens*, as St. Thomas [2] and the author of the *De natura generis* do.[3] At most, *res* is sometimes contrasted with *formalitas* to indicate what we might call a distinction between the *physical* and *metaphysical* order, both of which are *a parte rei*.

These notions, as well as the corresponding objective perfections which they signify, are not formally identical. From the viewpoint of concepts, they add to the notion of being certain formal aspects which are neither implicitly nor explicitly contained in the simple *ratio* " a being". From the standpoint of objective reality, they are formally distinct from one another and from the formal perfection signified by the quidditative concept, *a being*.[4] In the language of Philip the Chancellor, they are " three concomitant conditions " of being.[5] They are said to be convertible with being, not in the sense that they are

---

[1] *Summa theologica* I, n. 103; t. I, 162: Cum bonum dicatur dupliciter, honestum et utile ... Honestatem autem voco intelligibilem pulchritudinem.

[2] St. Thomas, *De Veritate*, q. 1, a. 1.

[3] *Opusculum de Natura Generis*, c. 2. Grabmann considers this an authentic work of St. Thomas; Mandonnet, however, rejects it.

[4] Confer chapter 2, p. 27, note 42.

[5] *Summa de bono*, prol. q. 7 (MSS Padoue Antonienne 156) f 3vb— Philip follows Avicenna. Confer the latter's *Metaphysica* (Venice, 1495) tr. V, c. 1, p. 35ab, where he speaks of the " conditiones concomitantes."

formally identical, but in the sense that "whatever is a being is good, and vice versa."[6] Or to put it in the language of Scotus, of whatever being is predicable *in quid,* truth, unity, and goodness are predicable *in quale.* In other words, in every physical entity or *res* the perfections expressed by the concepts being, one, true, good are all *unitive contenta* in one real indivisible whole.[7] The coextensive transcendentals are not primary notions or perfections, because as attributes they presuppose a subject which they modify or qualify. The primary subject of existence and consequently of further real modification is expressed by the quidditative concept of being. But while the concept of being as primary subject is a prerequisite for the very conception of an attribute like true, or one, it enters the definition, as has been pointed out, only extrinsically, that is, by way of addition (ἐκ προσθέσεως).[8] The concept of the attribute cannot be analyzed in terms of an intrinsic *quid* and a *quale,* that is, an intrinsic determinable and determining element. In its formal character as attribute, it is intrinsically only determining and as such irreducibly simple (*simpliciter simplex*).[9]

I. TRANSCENDENTAL UNITY

Numerical unity in so far as it is considered the basis of a multitude implies limitation and is reduced to the category of quantity.[10] Yet in a broader sense of unicity, numerical unity does not necessarily imply imperfection or limitation; otherwise

---

[6] Philip the Chancellor, *Summa de bono,* q. 1, f. 1vb (quoted from H. Pouillon, *op. cit.* p. 50): Bonum et ens convertuntur quia quidquid est ens est bonum et e converso.

[7] *Oxon.* 2, d. 16, q. un., n. 17; XIII, 43a: Isto modo ens continent multas passiones, quae non sunt res aliae ab ipso ente . . . distinguuntur tamen ab invicem formaliter et quidditative, et etiam ab ente, formalitate dico reali et quidditativa. *Ibid.* 3, d. 8, q. un., n. 17; XIV, 377b; *Metaphy.* 4, q. 2, n. 24; VII, 171b-172a.

[8] Chapter 4, pp. 88 ff.

[9] *Oxon.* 1, d. 3, q. 3, n. 12; IX, 110b: quilibet talis conceptus [sc. ultimarum differentiarum et passionum] est simpliciter simplex.

[10] *Metaph.* 4, q. 2, n. 18; VII, 167-168a.

it would be impossible to speak of the unicity of the Divine Nature. The question naturally arises, How does transcendental unity differ from predicamental unity? Scotus answers that they are not *really* distinct at all in creatures.[11] His position may be explained further by pointing out that the distinction between the two is that of a formal perfection and its mode. This distinction is not only characteristic of unity. It is common to all the so-called pure perfections or those whose formal *ratio* does not involve limitation.[12] Wisdom, for instance, is a quality in Solomon, yet it is substantial in God. Substance itself is a categorical notion as applied to creatures but not as applied to God. It is similar with such perfections as relations, active potency, or any others that are found in both God and creatures. As these perfections exist in creatures they are associated with certain modes which involve imperfection. Wisdom in man is associated with the modalities of contingency, finiteness, inaliety, and so on. By prescinding from these intrinsic modes it is possible to form a common univocal notion predicable of both God and creatures. As has been said, the concept which prescinds from the modes is imperfect and common; that which includes the modes is a proper notion.[13] Where the former is always univocal, the latter may be analogous. The difference between predicamental and transcendental unity is simply this. The former includes besides the univocal common transcendental notion the intrinsic mode of finiteness which immediately contracts it to the realm of the categories. Consequently, Scotus can say that the predicamental character of a perfection does not necessarily impede the abstraction of a common univocal

[11] *Ibid.* 167b: Tenendo ergo... quod in omni creato unitas convertibilis cum ente, realiter non differt ab unitate de genere quantitatis, licet semper, ut dictum est, conceptus unius transcendentis sit generalior; quia de ratione partis est possibilitas ad formam totius, et unum inquantum convertibile cum ente, non dicit limitationem, ideo ex tali inquantum tale, non fit numerus...

[12] *Oxon.* 1, d. 8, q. 3, n. 30; IX, 629b-630; *Metaph.* 4, q. 2, n. 11-12; VII, 163b-164.

[13] Chapter 2, pp. 25-27; Chapter 3, p. 56.

notion, transcendental in character.[14] The distinction between transcendental and predicamental unity, then, is merely modal. Consequently, Scotus can say that in creatures they do not differ really, that is, by a real distinction.

Unity, or better, "the one," may be described as that which is "undivided in itself and is divided from all else." [15] Though described in terms of negations, unity itself is a positive perfection and adds its own proper entity to being.[16]

Unity as described above is in reality an equivocal term. Each grade of being has its own proper unity.[17] Of whatever being is predicable *in quid,* unity is predicable *in quale.* But being is predicable quidditatively not only of individuals but also of the common elements which real beings possess. Scotus, as was indicated in the preceding chapter, tells us that being is included quidditatively in generic and specific as well as in individual notions.[18] Hence Scotus can speak of two principal types of unity, the unity of the singular or individual and the unity of the common nature.

Individual Unity as a Transcendental Attribute

The most perfect form of transcendental unity, and that which most truly deserves to be referred to as an attribute of being, is that unity or oneness known as "singularity." [19] It is the unity

---

[14] *Oxon.* 1, d. 8, q. 3, n. 30; IX, 629b-630: De sapientia dico quod non est species generis ut transfertur ad divina, nec secundum illam rationem transfertur, sed secundum rationem sapientiae, ut est transcendens; quomodo autem tale potest esse transcendens, dictum est superius.

[15] *Oxon.* 4, d. 6, q. 1, n. 4; XVI, 532a: Unum est in se indivisum, et ab alio divisum.

[16] *Metaph.* 4, q. 2, n. 13; VII, 165a; *Oxon.* 2, d. 3, q. 2, n. 4; XII, 80ab; *Ibid.,* q. 6, n. 11, 134b.

[17] *Oxon.* 2, d. 3, q. 4, n. 6; XII, 95a: Quamlibet enim entitatem consequitur propria unitas; *Ibid.* n. 20; 112a: omne ens secundum quamcumque entitatem consequitur propria unitas.

[18] Chapter 4, p. 78.

[19] *Oxon.* 1, d. 23, q. un., n. 2; X, 259a: Unitas est passio entis, sicut patet 4 *Metaph.* et per consequens consequitur rem ex natura rei. Et maxime est verum de illa unitate, quae est vera unitas, cujusmodi est

of the individual, or numerical unity.[20] Though not formally identical or synonymous, singularity is somehow coextensive with all real being. Whatever exists or can exist may be called numerically one. Numerical unity is not a mere mental fiction or second intention.[21] It is a real attribute in the sense that it expresses a formal perfection of the individual, that it is the terminus or *ratio objectiva* of a first intention, and that it is predicable of all real things *in quale*.

Individual unity is set in contrast to " unity of kind," that is, a unity existing between individuals by reason of a common nature. It is a unity designated properly by the pronoun "this ", the *unitas signata ut haec*. The irreducible simplicity (*simpliciter simplex*) of unity as a concept forbids any adequate explanation. It is recognized as implying a repugnance to being multiplied, or in scholastic language, " an impossibility of being divided into subjective parts." [22]

Closely connected with the notion of individual unity is that of the principle of individuation. The latter is regarded as the formal reason why the individual is incapable of being multiplied as such. It is the *ratio* or cause not of singularity in general, but

---

unitas individui. *Ibid.* 2, d. 3, q. 2, n. 4; XII, 80b: ...in unitate perfectissima, quae est unitas numeralis.

[20] *Oxon.* 2, d. 3, q. 4, n. 3; XII, 93a: Contra istam conclusionem arguo quatuor viis. Primo ex identitate rationis numeralis sive individuationis vel singularitatis.

[21] *Oxon.* 1, d. 23, q. un., n. 2; X, 259a: Unitas est passio entis, sicut patet 4 *Metaph.* et per consequens consequitur rem ex natura rei. Et maxime est verum de illa unitate, quae est vera unitas cujusmodi est unitas individui; ergo talis unitas non dicit intentionem secundam.

[22] *Oxon.* 2, d. 3, q. 4, n. 3; XII, 93a: Primo expono, quid intelligo per individuationem, sive unitatem numeralem, sive per singularitatem, non quidem unitatem indeterminatam, secundum quam quodlibet in specie dicitur unum numero, sed unitatem signatam ut hanc, ita quod sicut prius dictum est, quod individuum impossibile est dividi in partes subjectivas.—Subjective parts are contrasted to essential parts; the former refer to the order of extension, the latter to comprehension or intension. Thus all the genera, species or individuals that can be grouped together under one class or category are said to be the subjective parts of that class.

hujus singularitatis in speciali signatae, that is to say, in so far as something is determined to be just this.[23] It is a positive entity or formality over and above the specific nature possessed by each individual, and is referred to as *haecceitas*, though this designation is more Scotistic than Scotus'.[24] It is characteristic of all created natures, be they angelic or material, substantial or accidental.[25] Only infinite uncreated Being is of itself *haec natura*. Its pure actuality and fullness of perfection forbid any further determination or multiplication in kind. All created natures, or finite natures *qua* natures, have no intrinsic repugnance to being duplicated, triplicated and so on. They require a determinant formally extrinsic to the nature *qua* nature by which they are rendered incapable of further multiplication within the same species.[26] The *haecceitas* does just this. It is related to the common nature as a specific difference is related to the generic perfection, but where the former produces a difference in kind the latter gives rise to a difference of individuals.[27] It is only

---

[23] *Ibid.*; confer also *ibid.* q. 2, n. 4; 80a.

[24] Scotus does not use this designation in his principal work, the *Opus Oxoniense*. We find only a few references to the term in the *Rep. Par.* 2, d. 12, q. 5, n. 1; XXIII, 25b; *Ibid.* n. 8; 29a; 31b; n. 12; 32a and n. 13.

[25] *Oxon.* 2, d. 3, q. 7, n. 6; XII, 167a: Per idem patet ad Avicennam; dico quod intentio ejus fuit quod tantum sit unus Angelus in una specie, sed propositio cui haec conclusio innititur, scilicet quod Angelus superior creat inferiorem, a nullo Theologo vel Catholico conceditur, quare nec ejus conclusio debet concedi ab aliquo Theologo. *Ibid.* qq. 1-7, passim Confer also following note.

[26] *Ibid.* q. 7, n. 5; 161b-162a: Dico igitur, quod omnis natura, quae non est de se actus purus, potest secundum illam realitatem, secundum quam est natura, esse potentialis ad realitatem illam qua est haec natura; et sicut de se non includit aliquam entitatem quasi singularem, ita non repugnant sibi quocumque tales entitates, et ita potest in quotcumque talibus inveniri. In eo tamen quod est necesse esse ex se, est determinatio in natura ad esse hoc, quia tantum est, quantum potest esse, quia quidquid potest esse in natura, est ibi, ita quod determinatio non potest esse per aliquod extrinsecum ad singularitatem, si possibilitas sit in natura per se ad infinitatem, secus est in omni natura possibili, ut potest cadere ['eadem' in older editions] multitudo.

[27] *Ibid.* q. 4, n. 20, 112b: Non ... est intra formalem rationem quidditatis ut quidditas est, sed est quasi passio consequens quidditatem, et omne tale apud eum [sc. Avicennam] vocatur accidens. Et hoc modo Philosophus

formally distinct, but really identical with the common nature.[28]

Because the *haecceitas* is an entity or formality, it adds to the total entity of the being. As a result each individual has a distinct grade of being, a being proper to itself. The individual is an *ens per se.* Hence it is also an *unum per se.* This unity which characterizes the individual is what Scotus understands by individual unity as a transcendental attribute of real being. It is not, therefore, simply the *haecceitas,* but it is a property or attribute of the metaphysical composite, namely, the common nature plus the individuating difference or *haecceitas.* The *haecceitas* is characteristic only of created natures and explains why in the world of real existence they are always individualized and incapable of multiplication. Singularity as a universal attribute of being is broader. It is a property of any being incapable of further multiplication in kind, whether this repugnance to multiplication be due to the very nature itself, as is

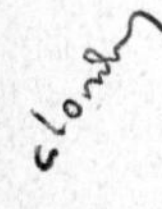

the case with the divine nature, or whether it be due to a distinct formal principle of individuation, as is the case with all finite natures.

Whatever exists or can exist, therefore, may be called numerically one. In God this numerical unity is a characteristic of his nature as nature. In created individuals, however, it is possible to distinguish between two formalities, the *haecceitas* and the common nature. The individual Peter, for instance, possesses the essential nature of a man (*natura communis*) plus an element which is unique and proper to himself (*haecceitas*). How then are created natures said to be numerically one? The *haecceitas* or formal principle of individuation is primarily

---

quandoque accipit accidens, a quo dicitur fallacia Accidentis, pro omni eo quod est extra rationem formalem alterius; omne autem tale est extraneum illi alteri, ex comparatione ad aliud, et hoc modo differentia accidit generi, et quidquid est individuans, accidit naturae specificae, sed non sicut ipsi intelligunt de accidente, et ideo ibi est aequivocatio de *accidente. Ibid.* q. 6, n. 2; 128a: ergo aliquid per se includitur in ratione individui, quod non includitur in ratione naturae. Illud autem inclusum [sc. haecceitas] est entitas positiva ... et facit unum per se cum natura ...; ergo est per se determinans illam naturam ad singularitatem, sive ad rationem illius inferioris. *Ibid.* n. 9; 133a.

[28] *Oxon.* 2, d. 3, q. 6, n. 15; XII, 144ab.

numerically one; the individual (Peter) is *per se* numerically one; the common nature (man) is *denominative* numerically one, in the same way that animal in man can *denominative* be called rational.[29]

Unity of the Common Nature

Another form of unity which to a certain extent is transcendental in character is the unity of the common nature. Scotus refers to it as a " unity that is less than numerical unity." [30]

The *natura communis* has a certain indifference about it. There is no intrinsic or formal reason to be found in a created nature as nature why it should be one or why it should be many, why it need be universal or particular.[31] It is aloof, as it were, and prior by nature to any determination. Human nature, for instance, is not precisely or formally Winston Churchill; neither is it the human race as a whole (logical *totum*); it is not merely a concept predicable of many (the formal universal), for as such it can exist only in a thinking mind. On the other hand, it is not formally singular or individual for then it could not be predicated of the various members of the genus *Homo*. True, human nature is never found either in thought or in reality except in one or the other of these ways. Nevertheless, in its formal character as *natura humana* it is indifferent to any one of them. Avicenna expressed this when he said: *equinitas sit tantum equinitas, nec ex se una, nec plures, nec universalis, nec particularis.*[32]

---

[29] *Ibid.* n. 10; 134a: Ita concedo quod quidquid est in isto lapide, est unum numero, vel primo, vel per se, vel denominative. Primo forte ut illud, per quod unitas talis convenit huic composito; per se, ut hic lapis, cujus illud quod est primo unum hac unitate, est per se pars; denominative tantum, ut illud potentiale quod perficitur isto actuali, quod quasi denominative respicit actualitatem ejus et unitatem similiter.

[30] *Oxon.* 2, d. 3, q. 1, n. 2; XII, 7b-8a; confer also the following note.

[31] *Ibid.* n: 7; 48a: Aliqua est unitas in re realis absque omni operatione intellectus, minor unitate numerali, sive unitate propria singularis, quae unitas est naturae secundum se; et secundum istam unitatem propriam naturae, ut natura est, natura est indifferens ad unitatem singularem; non ergo de se est sic una unitate illa, scilicet unitate singularitatis.

[32] *Ibid.* '48ab: Qualiter autem potest hoc intelligi, potest aliqualiter videri per dictum Avicennae 5 *Metaph.* ubi vult, quod 'equinitas sit

Commonness is not to be understood here in the sense of numerical identity, as the divine essence is said to be common to the Three Divine Persons.[33] It is a commonness in kind. It refers ordinarily to the specific nature, though it need not be restricted to the species. We can as readily speak of a common generic nature, for instance, substance, or even of a common transcendental nature, for example, being. This common, or even better, this indifferent nature realized in countless individuals is the objective basis of the universal concept. It is not of itself formally universal.[34] Hence it is a complete misunder-

---

tantum equinitas, nec ex se una, nec plures, nec universalis, nec particularis.' Intellige, non est ex se una unitate numerali; nec plures, pluralitate opposita illi unitati, nec universalis actu, eo modo quo aliquid est universale factum ab intellectu, non ut objectum intellectus, nec est particularis de se; licet enim nunquam sit realiter sine aliquo istorum, non tamen est de se aliquod istorum, sed est prius naturaliter omnibus istis.

[33] *Ibid.* n. 9; 54-55.

[34] *Ibid.* n. 8; 54ab: Ad primum dico, quod universale in actu est illud quod habet unitatem indifferentem, secundum quam ipsum idem est in potentia proxima, ut dicatur de quolibet supposito, quia *primo Posteriorum:* 'universale est, quod est unum in multis et de multis.' Nihil enim secundum quamlibet unitatem in re est tale, quod secundum ipsam unitatem praecisam sit in potentia proxima ad quodlibet suppositum, ut dicatur de quolibet supposito praedicatione dicente: *hoc est hoc,* quia licet alicui existenti in re non repugnet esse in alia singularitate ab illa, in qua est, non tamen illud vere dici potest de quolibet inferiori, quod quodlibet est ipsum, hoc est enim solum possibile de objecto eodem indifferenti actu considerato ab intellectu, quod quidem ut intellectum habet unitatem etiam numeralem objecti, secundum quam ipsum idem est praedicabile de omni singulari dicendo quod hoc est hoc. Ex hoc apparet improbatio illius dicti quod intellectus agens facit universalitatem in rebus, per hoc quod denudat ipsum *quod quid est* in phantasmate existens, nam ubicumque est, antequam in intellectu possibili habeat *esse* objective, sive in re sive in phantasmate, sive habeat esse certum, sive deductum per rationem, et si sit non per aliquod lumen, sed semper sit talis natura ex se, cui non repugnet esse in alio; non tamen est tale, cui potentia proxima convenit dici de quolibet, sed tantum est in potentia proxima, ut est in intellectu possibili; est ergo in re commune, quod non est de se hoc, et per consequens ei de se non repugnat esse non hoc. Sed tale commune non est universale in actu, quia deficit ei illa differentia, secundum quam completive universale est universale, secundum quam scilicet ipsum idem aliqua identitate est praedicabile de quolibet individuo, ita quodlibet sit ipsum.

standing of Scotus to say that he maintains the existence of the universal *a parte rei.* Commonness does not mean simply universality.[35] It expresses rather an indifference to either universality or singularity. Formal universality is a second intention.[36] It is a relation that the *natura communis* takes on by reason of the fact that it exists in the mind. The *natura communis* as such, that is, in its state of formal indifference to either singularity or universality, is the object of a first intention. It is *per se* the object of the intellect; as such it is considered by the metaphysician and expressed in a real definition. From the standpoint of propositions, it is the subject of all essential (*per se primo modo*) predication.[37]

The fact that this nature, though formally distinct, is really identified with the *haecceitas,* without which it is incapable of any extra-mental existence or operation, does not destroy its formal indifference. When we grasp just this aspect or part of

---

[35] *Ibid.,* n. 10, 55b: Communitas convenit naturae extra intellectum, et similiter singularitas. Et communitas convenit ex se naturae, singularitas autem convenit naturae per aliquid in re contrahens ipsam; sed universalitas non convenit rei ex se, et ideo concedo, quod quaerenda est causa universalitatis, non tamen quaerenda est causa communitatis alia ab ipsa natura; et posita communitate in ipsa natura secundum propriam entitatem et unitatem, necessario oportet quaerere causam singularitatis, quae superaddit aliquid illi naturae cujus est.

[36] *Metaph.* 7, q. 18, n. 6; VII, 456b: Sumitur enim [sc. universale] quoniam pro intentione secunda quae scilicet est quaedam relatio rationis in praedicabili, ad illud de quod est praedicabile, et hunc respectum significat hoc nomen *universale* in concreto, sicut et *universalitas* in abstracto.

[37] *Oxon.* 2, d. 3, q. 1, n. 7; XII, 48b: Et secundum istam prioritatem naturalem est *quod quid est,* et per se objectum intellectus, et per se ut sic, consideratur a Metaphysico, et exprimitur per definitionem; et propositiones verae primo modo sunt verae ratione quidditatis sic acceptae, quia nihil dicitur de quidditate per se primo modo, nisi quod includitur in ea essentialiter, inquantum ipsa abstrahitur ab omnibus istis, quae sunt posteriora ipsa naturaliter. Non solum autem ipsa natura est de se indifferens ad esse in intellectu, et in particulari, ac per hoc ad esse universale et singulare, sed et ipso habens *esse* in intellectu, non habet primo ex se universalitatem; licet enim ipsa intelligatur sub universalitate, ut sub modo intelligendi ipsam, tamen universalitas non est pars conceptus ejus primi, quia non conceptus Metaphysici, sed Logici. Logicus enim considerat secundas intentiones applicatas primis, secundum ipsum Avicennam.

reality, we see no reason why it should be one rather than many. This nature has a sort of oneness, a unity of its own—the unity of the nature as such. This unity is not the same as singularity or individual unity, since no finite nature as such is singular. It is therefore a unity that somehow is less than numerical unity, yet is real for all that. For like the common nature itself, this unity exists *a parte rei* and is not projected into reality by the thinking mind.

Though unity belongs to the nature as common or indifferent to either one or many, it is rather an attribute of the common nature than formally identical with it.[38] But where individual unity is properly speaking an attribute of the individual, and from a logical standpoint belongs only *per accidens* to the nature *qua* nature, this unity of kind is a property of the nature *qua* nature and is predicable of it *per se secundo modo.*

This theory of the unity of the indifferent or common nature has opened Scotus to the charge of exaggerated realism. Disregarding Scotus' own explanation, interpreters have accused Scotus of advocating some sort of real numerical identity between the "common" nature of individuals. It is easy to see how this interpretation when applied to his doctrine of the common concept of being, could even lead to the charge of monism and pantheism. Modern investigation and study of Scotus' actual doctrine have exploded these accusations. Minges[39] and Kraus,[40] for example, have shown that this unity, like the com-

---

[38] *Ibid.*, 49a: Hoc ergo modo intelligo naturam habere unitatem realem minorem unitate numerali, et licet non habeat eam de se, ita quod sit intra rationem naturae, quia 'equinitas est tantum equinitas' secundum Avicennam . . . tamen illa unitas est passio propria naturae secundum suam entitatem primam, et per consequens nec est ex se haec intranee, nec secundum entitatem propriam necessario inclusam in ipsa natura secundum primam entitatem suam.

[39] P. Minges, "Der angebliche exzessive Realismus des Duns Scotus" in *Beiträge zur Geschichte der Philosophie des Mittelalters,* VII (Münster, 1908).

[40] J. Kraus, *Die Lehre des Johannes Duns Skotus von der Natura Communis* (Freiburg, Studia Friburgensia, 1927), p. 136: Der in dieser Weise indifferenten Natur folgt eine eigene, ihr genügende reale Einheit, die geringer ist als die numerische; es ist die Einheit der Natur als solcher.

mon nature of which it is the property, is not numerically identical in all individuals of the same species, but differs with each individual being that actually exists. As Scotus himself insists, this *unitas realis non est alicujus entitatis existentis in duobus individuis sed in uno.*[41]

As the common nature is actually found in concrete beings the definition of unity is verified of it in its formal character as nature. It is undivided in itself and divided from all else.

II. TRANSCENDENTAL TRUTH

Truth, or better, "true," is a transcendental property of all beings. It is predicable therefore of being itself and of all of which being can be predicated.[42] Like unity, goodness or the attributes in general, it is predicable *in quale* whereas being is predicable *in quid.*[43]

Wie diese selbst, ist auch die spezifische Einheit nicht in allen Individuen derselben Art numerisch dieselbe, sondern in den einzelnen existerenden Dingen real verschieden. Doch ist sie mehr als eine Einheit der Ähnlichkeit; denn sie bildet deren Grundlage und Fundament. It should be noted, however, that Kraus' explanation of the *natura communis* is defective on several counts. For one thing, as we have mentioned already, he fails to grasp the significance of the formal distinction in this connection and as a result creates the impression that the *natura communis* is a sort of physical entity pertaining to the *physical* rather than the *metaphysical* order. Gilson seems to have been influenced by Kraus in the absurd interpretation he gives of the difference between the teaching of Scotus and that of St. Thomas regarding the being which is the term of the creative act. Confer his "Les sieze premiers Theoremata et la Pensée de Duns Scot" in *Archives d'Histoire Doctrinale et Littéraire du Moyen Age,* XII-XIII (1937-1938), p. 77. For a further criticism of Kraus' interpretation of the *natura communis,* see P. Boehner, "Scotus' Teaching According to Ockham: II. On the *Natura Communis*", *Franciscan Studies,* VI (1946), 395-407.

[41] *Oxon.* 2, d. 3, q. 6, n. 10; XII, 133b: In eodem ergo quod est idem numero, est aliqua entitas, quam consequitur minor unitas quam sit unitas numeralis, et est realis, et illud cujus est unitas talis formaliter, non est de se unum unitate numerali. Concedo ergo, quod unitas realis non est alicujus entitatis existentis in duobus individuis, sed in uno.

[42] *Oxon.* 1, d. 3, q. 3, n. 20; IX, 145b: *Verum* est passio entis et cujuslibet inferioris ad ens.

[43] *Ibid.: Verum* ... non dicitur *in quid* de ente, nec de aliquo per se inferiora ad ens. *Oxon.* 1, d. 8, q. 3, n. 23; IX, 616b: Aliqua praedicata transcendentia dici *in quale,* ut *verum.*

It is not easy to give a clear-cut notion of truth since the term has a variety of meanings. Wilpert has called attention to the change of emphasis in Aristotle's conception of truth. In his earlier writings under the influence of Plato's spirit he speaks of truth primarily in an ontological sense. In his later works it is restricted almost exclusively to the logical sphere.[44] Nevertheless, the Platonic tradition not only survived but through the efforts of St. Augustine became inseparably associated with the Christian exposition of creation, the divine ideas and the theological explanations of the eternal generation of God the Son by God the Father. As a result, though the primary meaning of truth referred to a property of mental propositions (*veritas in intellectu*), the medieval schoolmen devoted a great deal of attention to *veritas in rebus.*

In the *Quaestiones super libros Metaphysicorum Aristotelis* Scotus devotes an entire question to the discussion of truth. He attempts to clarify the various meanings of *veritas in re* that were in common use by the thirteenth century philosophers.

Things are said to be true in either one of two general ways: 1) by reason of their conformity to that which produced them, or 2) by reason of their conformity to an intellect which knows them.[45]

Though he does not mention it expressly, the common faculty or power he has in mind in both instances is the intellect. The intellect in general, like the divine intellect in particular, is not simply a passive faculty but an active, creative, productive power. In consequence, even the causality of the exemplar cause takes on a new significance in the system of Duns Scotus.

Truth as Conformity to a Productive Mind

Even truth as a conformity to a productive intellect is an equivocal term. It may express a simple conformity, understanding the latter in a general or unqualified sense. Or it may

[44] Paul Wilpert, "Zum aristotelischen Wahrheitsbegriff" in *Philosophisches Jahrbuch,* LIII (1940), 3-16.

[45] *Metaph.* 6, q. 3, n. 5; VII, 337a: Est enim veritas in rebus et veritas in intellectu. In rebus dupliciter, in genere videlicet per comparationem ad producentem, et per comparationem ad cognoscentem sive intelligentem.

be further specified as a conformity of adequation or perfect similarity. Or thirdly, as a conformity of imperfect similarity or imitation.[46]

The first member of the trio is in a sense common to the other two. The second is perfectly exemplified in the case of the Son of God. For, as St. Augustine points out, the Second Person of the Blessed Trinity is called " Truth " because He is equal to the Father who begets Him in an eternal generation. Here the highest possible degree of similarity exists between the *principium* and the *principiatum*, between the *Pater dicens* and the *Verbum dictum*. An example of the third is to be found in creatures, which are imitations of the divine ideas.[47] It is not our purpose to discuss here how Scotus' exemplarism differs from that of the other scholastics. We merely note in passing that Scotus does not simply identify the ideas with the divine essence as such. He regards them rather as artistic productions of an infinitely fertile intellect.[48]

Truth as a Relation to a Knowing Mind

Here again a thing may be said to be true in three ways. The first and most important is by reason of its very being or *esse*. Every being, by the very fact that it is a being, has an inherent capacity or aptitude to manifest itself to an intellect capable of perceiving it. In this sense truth is simply *esse intel-*

---

[46] *Ibid.:* Primo modo dicitur veritas absolute conformitas producti ad producens, aut determinate conformitas talis secundum adaequationem, aut determinate conformitas secundum imitationem.

[47] *Ibid.:* Et licet primus istorum trium modorum videatur esse communis secundo et tertio, tamen si nomen *veri* imponatur ad significandum quodcumque trium praedictorum secundum propriam rationem, erit aequivocum. Secundus modus invenitur in Filio Dei, qui veritas est, quia est secundum Augustinum summa similitudo principii, haec enim est conformitas cum adaequatione. Tertius modus invenitur in creatura, quae imitatur exemplar, cui aliquo modo assimilatur, defective tamen, alias non diceretur imitari.

[48] Confer for instance: *Oxon.* 1, d. 35, q. un., X, 536-560; *Ibid.* d. 36, q. un.; 564-587; *Ibid.* d. 43, q. un.; 728-784; *Oxon.* 2, d. 1, q. 1; XI, 6-47; etc.

ligible or *esse ut manifestativum sui.* Secondly, we may speak of truth in things in so far as they play an active role in the cognitive process. In so far as the object acts as a partial cause in the process of assimilation that we know as cognition, the thing or object is said to have truth in itself. As Scotus puts it, the thing is *assimilativa intellectus assimilabilis.* That this *esse assimilativum* is not identical with the *esse manifestativum* becomes clear if one compares the role of "object" in the divine knowledge and in human cognition. In theological language, the divine essence, the primary object of the divine intellection, does not determine but merely terminates God's act of knowledge. For the divine intellect is in no sense passive. The divine essence therefore has no *esse assimilativum* or active aspect in regard to it, though the divine essence is by its very nature infinitely intelligible, and therefore possesses *esse manifestativum* in an infinite degree. The *esse assimilativum,* then, implies a relation to an intellect in some sense passive, or as Scotus put it, it is characteristic of created intelligence only.[49]

This peculiar distinction is of importance, for it reveals that Scotus did not merely attribute a terminative causality to the object in ideogenesis. In opposition to Henry of Ghent, Peter Olivi, Peter de Trabibus and William of Nottingham, he maintains the causal influence of the object on the cognitive faculty,[50] though not to the same degree as did Aristotle.[51] For Scotus

---

[49] *Metaph.* 7, q. 3, n. 5; VII, 337ab: Secundo modo, scilicet per comparationem ad intellectum dicitur res vera tripliciter. Primo, quia sui manifestativa, quantum est de se cuicumque intellectui potenti manifestationem cognoscere. Secundo, quia assimilativa intellectus assimilabilis, qui non est nisi intellectus creatus. Tertio, quia facta manifestatione vel assimilatione, res in intellectu est, sicut cognitum in cognoscente.... Esse autem assimilativum dicit rationem activi respectu assimilabilis, et sequitur naturaliter esse manifestativum, vel disparatum est non habens ordinem ad ipsum, sed semper assimilativum, et assimilatio respectu intellectus passivi praecedit hoc quod est esse in intellectu.

[50] Confer B. Jansen, *Die Erkenntnislehre Olivis* (Berlin, 1921) 48-63; 60-61.

[51] E. Longpré, "The Psychology of Duns Scotus and its Modernity" in *Franciscan Educational Conference,* XIII (Nov. 1931), pp. 15-77.

retains much of the Augustinian tradition of the essentially active character of the intellect.[52]

In a third sense philosophers are accustomed to attribute truth to a thing in so far as it is said to be in the intellect *sicut cognitum in cognoscente.* In this sense the "thing" is simply the content of a concept, an *ens diminutum.* In its character as an *esse objectivum,* it is set over against the knowing intellect, as it were. To this extent one can speak of a corresponding relation between the thought of the mind (act of intellection) and the content of the thought. This relation is obviously only an *ens rationis,* since the thought content itself is only mentally distinct from the thought. In this third sense "truth" obviously does not belong to metaphysics properly but to logic.[53]

Metaphysical Truth

This distinction of Scotus between truth as a relation to a productive principle and to a knowing mind is interesting. Platonism, and to a lesser extent Avicennian-Augustinian illuminationism, tend to confuse the two orders in the sense of seeking the *ratio cognoscendi* of objects in the pure intelligibility of ideas existing apart from the material sensible world, or even of making the intelligibility of the object depend somehow upon its conformity to the divine intellect. Among the propositions condemned by Bishop Tempier in 1277 were two concerning an ontologistic interpretation given to Augustinian exemplarism

[52] *Ibid.,* confer also H. Klug, "L'activité intellectuelle de l'âme selon le B. Duns Scot" *Études Franciscaines,* XLI (1929), 517-520. The importance of the doctrine of partial causality in this connection has been emphasized by R. Messner in *Schauendes und begriffliches Erkennen nach Duns Skotus* (Freiburg, Herder, 1942).

[53] *Metaph.* 7, q. 3, n. 15; VII, 346a: Tertium est ens diminutum, et est ens Logicum proprie, unde omnes intentiones secundae de tali ente praedicantur, et ideo proprie excluditur a Metaphysico. Convertitur tamen cum ente aliqualiter, quia Logicus considerat omnia ut Metaphysicus; sed modus alius considerationis, scilicet per quid reale et per intentionem secundam, sicut convertibilitas entis simpliciter et diminuti, quia neutrum alterum excedit in communitate, quidquid enim est simpliciter ens, potest esse ens diminutum.

(propositions 8 and 9).[54] Perhaps the remembrance of this was still fresh in Duns Scotus' mind when with one stroke he separates the two orders in such a way as to prevent any error of ontologism at the outset. The order of intelligibility (*esse manifestativum*) and of causality on the part of the object (*esse assimilativum*) are radically distinct from any relations the object may have to the divine intellect. This is true says Scotus even though it be the same individual which produces the thing and has knowledge of it. So emphatic is he on this point that he insists that if *per impossibile* God were to produce something similar to Himself or in imitation of Himself and still would be incapable of knowing anything the first three types of truth would still exist without the second, and vice versa if there were a God who possessed intelligence but did not generate or produce things in being.[55]

If we ask, which of these six ways in which truth may be said to be in things properly pertains to the province of metaphysics, Scotus answers that the first three certainly do. For they imply a real conformity (either absolute or determined, either adequate or in imitation) of the product to that which produced it, and in this sense truth is a real perfection of the thing in question. Since it does not pertain properly to either physics or mathematics, it must fall under the investigations of the metaphysician. Truth in the sense of conformity to a knowing intellect must be distinguished. Only the first of the three members (*esse manifestativum sui*) pertains to metaphysics, for it is simply convertible or coextensive with being. The *esse assimilativum* is not convertible but restricted to the physical order.[56]

---

[54] M. DeWulf, *History of Medieval Philosophy,* trans. E. C. Messenger (New York, Longmans, Green & Co., 1938) II, 227.

[55] *Metaph.* 7, q. 3, n. 5; VII, 337b.

[56] *Ibid.* n. 15; 345b-346a: Secundum veritatem autem dico, quod ad considerationem Metaphysici pertinet verum reale primo modo sumptum, scilicet per comparationem ad producentem, et hic quoad omnia tria membra, quia non contrahunt ens ad quantum, nec ad motum. Secundum autem scilicet sumptum per comparationem ad cognoscentem ad primum membrum pertinet, quia convertitur cum ente. Secundum contrahit ad actum determinatum, nec convertitur cum ente. Tertium est ens diminutum, et est ens Logicum.

Truth as a Coextensive Attribute

While Scotus does not rule out conformity of created objects to the divine intellect or to the exemplar ideas from the consideration of the metaphysician, nevertheless this truth is not properly a *passio entis.* Only in the sense of intelligibility or the *esse manifestativum sui* is truth a coextensive attribute of being.[57] Far from depending on an act of the intellect, this truth is independent even of the actual existence of an intellect. For, Scotus says, in the hypothesis that no intellect existed, there would still be an inherent aptitude in every being to reveal or manifest itself according to its proper grade of being.[58] In this sense it may be said to be formally in beings themselves and is not derived, as it were, from the operation of the intellect. Consequently, even abstracting from the act by which God knows Himself, one can speak of the divine essence as possessing truth.

In an enlightening passage of the Oxford *Commentary* Scotus argues for a distinct objectivity of truth prior to the act of the mind. The mind (the divine intellect included) tends towards its respective object *sub ratione veri.* Consequently, this intelligibility is a condition that precedes by nature the act of the mind and cannot be derived from that act in the sense that the intellect confers upon being the superadded notion of truth. Were such the case, the intelligibility of being would be merely a *relatio rationis.* And since no such relation could be infinite, one could not speak of the *Ipsum Esse* as Infinite Truth.[59]

---

[57] *Ibid.:* Secundum autem sc. sumptum per comparationem ad cognoscentem [esse manifestativum] pertinet [sc. ad Metaphysicam], quia convertitur cum ente.

[58] *Ibid.* n. 5; 337b: Si nullus esset intellectus, adhuc quaelibet res secundum gradum suae entitatis, esset nata se manifestare; et haec notitia est, qua res dicitur nota naturae, non quia natura cognoscat illam, sed quia propter manifestationem majorem vel minorem nata esset, quantum est de se, perfectius vel minus perfecte, cognosci.—Compare this with St. Thomas, *De Veritate,* q. 1, a. 2c.

[59] *Oxon.* 1, d. 8, q. 4, n. 14; IX, 654a: Quaero igitur utrum veritas dicat praecise illam perfectionem, quae est in re formaliter, aut praecise illam relationem factam ab intellectu, aut utrumque? Si praecise relationem

In short, Scotus narrows down the field of truth as a coextensive transcendental attribute of being far more than does St. Thomas, or the neo-scholastics in general.[60] With the latter the conformity of things to the exemplar ideas of God plays a primary role. The basic reason for this difference is found in the fact that Scotus requires that the attributes, like the concept of being itself, be univocal. Since the conformity of the *exemplata* to the divine ideas is understood in a totally different sense from the conformity of the divine essence to the divine intellect, the two fundamentally different relationships cannot be brought together in a concept that is both real and univocal.

Though truth in the sense of intelligibility is a necessary condition for actual knowledge, *verum* is not formally the primary object of the intellect.[61] That is to say, the first thing the mind grasps is not the intelligibility, which is a relation, but the absolute entity (being) of which intelligibility is an attribute.

---

rationis; ergo non est perfectio simpliciter, quia nulla relatio rationis potest esse infinita.... Si ambo, cum illa non sint unum nisi per accidens, quia relatio rationis cum ente reali nunquam facit unum per se; quod patet, qui multo minus facit unum cum ente reali ens rationis quam passio cum subjecto; (passio enim consequitur subjectum ex ratione subjecti) nullum autem ens rationis consequitur ens reale ex ratione sui. Separata igitur ista duo, [nempe veritas et bonitas] quae concurrunt in isto ente per accidens, et sequitur tunc, quod veritas semper dicat praecise illam perfectionem in re, et bonitas similiter; et tunc ultra, cum nulla sit distinctio in re, sive secundum opinionem sive secundum expositionem opinantium, quia dicerent eamdem perfectionem, ut perfectio in re est, ut probatum est, et sine omni distinctione rei et rationis, sequitur quod bonitas et veritas sint formaliter synonyma, quod ipsi negant. *Ibid.* n. 11, 652b: Si ab aeterno Deus ex sui immaterialitate intelligit et vult se, et hoc sub ratione boni et veri; ergo ibi est distinctio veri et boni, rationum formalium in objectis ante omnem actum circa talia objecta.

[60] Confer, Raymond J. McCall, "St. Thomas on Ontological Truth" in *New Scholasticism,* XII (1938) pp. 1-29; Amato Masnovo, *Problemi di Metafisica e di Criteriologia* (Milan, 1930), esp. pp. 25-27; Mercier, *Metaphysique Generale* (Paris, 1910), pp. 184ff; Gredt, *Elementa Philosophiae Aristotelico-Thomisticae,* (Freiburg, 1937), II, 19-23.

[61] *Oxon.* 1, d 3, q. 3, n. 20; IX, 145ff.

III. TRANSCENDENTAL GOODNESS

Transcendental goodness as a convertible attribute of being represents a real perfection, formally distinct from truth and unity as well as from the perfection signified by the quidditative concept of being. Nevertheless, it is inseparably connected with these perfections by a real identity in the unity of a physical nature.

The univocal character of the transcendentals causes Scotus to distinguish here, as he did with the attribute of *true*, between goodness in the strict sense as coextensive with being and all analogical meanings of the term. All accidental, extrinsic or moral goodness, consequently, is excluded as a form or type of transcendental goodness. This, of course, does not preclude such forms of goodness, in so far as they possess a real or positive entity, from possessing a corresponding transcendental goodness or *bonitas naturalis*, as Scotus calls it.[62]

When divorced of all that accrues to it by way of attribution, the notion of transcendental goodness as a universal property of being loses much of its significance, particularly when understood in its original Platonic meaning. It is not surprising, then, to find in Scotus little more than passing mention of this attribute.

A word regarding the Platonic notion of goodness as it appeared in medieval speculation is helpful in understanding Scotus. The whole treatment of ontological goodness in the Middle Ages was profoundly influenced by Platonism. As Hirschberger put it, "Es ist nicht Aristoteles der Sein und Wert konfundiert hat, sondern Platon." [63] What he has shown to be the basic meaning of transcendental goodness in the philosophy of Albert the Great

[62] *Oxon.* 2, d. 7, q. un., n. 11; XII, 386ab: Dico quod ultra bonitatem naturalem volitionis, quae competit sibi inquantum est ens positivum, quae etiam competit cuicumque enti positivo, secundum gradum suae entitatis magis et minus, praeter illam est triplex bonitas moralis....

[63] Hirschberger, "Omne ens est bonum" in *Philosophisches Jahrbuch,* LIII (1940), 298.

and of St. Thomas in particular can be extended to the scholastics in general, at least up to the time of Scotus.[64]

Ontologic goodness is not primarily the appetibility of a being or its ability to perfect something beyond itself. It is rather the actualization of an ideal within the being itself. To the extent that something is actual and possesses a certain grade of being, it has realized perfection. It is primarily good, as Scotus put it, because it has "perfection in itself and in reference to itself."[65]

The notion of a goal or τέλος is fundamental in the Platonic notion of the good. Not only do rational creatures strive for a goal but irrational beings have a goal-ward tendency. Hence the importance of the *appetitus naturalis* for the schoolmen. Their *natura*, like the Greek's φύσις, is a principle of activity, of striving, of operation. It possesses within itself the potentiality to become that for which it strives. Similarly, what it has actually attained, that is to say, what it actually is, can be regarded as the fulfilment of a goal or as the actualization of this potentiality.

It is not surprising, then, to find Philip the Chancellor[66] or Alexander of Hales[67] defining good as "the indivision of act and potency, where act is regarded as the complement or the perfecting of a possibility of which the being is capable by nature." Here the Platonic notion is translated into Aristotelian termi-

---

[64] With Ockham the break with Platonism is complete. Ontologic goodness is simply the appetibility of a thing. *Summa Logica*, pars I, c. 10; Bonum etiam, quod est convertibile cum ente, significat idem quod haec oratio: aliquidem secundum rectam rationem volibile et diligibile.

[65] *Rep. Par.* 2, d. 34, q. un., n. 3; XXIII, 170b: Primum bonum dicit perfectionem in se et ad se.

[66] *Summa de Bono*, prol. q. 7. Philip refers the definition to Aristotle; Albert the Great attributes it to Avicenna: "Avicenna autem dicit in Metaphysica sua quod bonum est indivisio actus a potentia." *Quaestiones de bono* (*Summa de Bono*, q. 1-10), ed. H. Kühle, in *Florilegium Patristicum*, fasc. xxxvi (Bonn, 1933) q. 1, p. 10, line 12.

[67] *Summa theologica* I, n. 88, (t. 1, 140a): Unde bonum dicitur indivisio actus a potentia, et actus dicitur complementum sive perfectio possibilitatis, ad quam res nata est.

nology. In a word, things are said to be ontologically good in so far as their possibilities are actualized. Potentiality and actuality are no longer separated but have become one. They are what they can be. This is true of a being, no matter what stage of perfection it has reached. Provided it has any being or reality at all, it is what it is and to that extent has realized a goal and is a good. Hence the dictum met with in Philip[68] and attributed by Alexander of Hales to Boethius:[69] *bonum et ens convertuntur.* The fundamental meaning is not *bonum alteri* but simply *bonum sibi.* Only because of the primary goodness are things appetible to others.

God too can be called good in this sense, at least where this possibility or *potentia* is understood in the sense of logical possibility, that is, compatibility of notes or perfections. For God is all that He can be. He is the fulness of actuality. There is no further striving for perfection. He is the ἱκανόν, the τέλος μονοειδές, the ἀνυπόθετον of Plato. He is the absolute end in the hierarchical arrangement of relative goods, from whom all goodness and being have come.

With this general notion in mind, we turn to Scotus. Though he speaks of the good as a universal attribute of being more or less in passing, we do find an occasional passage that is more detailed.[70] For instance, the following occurs in the *Reportata Parisiensia:*

> I say that the *good* and the *perfect* are the same according to [Aristotle's] Metaphysics, book 6. The *perfect* has a twofold meaning. In one sense it means that to which nothing is wanting intrinsically. Such a thing is said to be perfect by reason of an essential intrinsic or primary perfection. In another way a thing is said to be perfect by reason

[68] Philip, *op. cit.*, q. 1, f. 1rb: Bonum et ens convertuntur quia quidquid est ens est bonum et e converso. (Quoted from Pouillon, *op. cit.*, p. 50, note 43.)

[69] Alexander of Hales, *ibid.* It is attributed by Alexander to Boethius. Confer the latter's *De hebdomadibus* (PL. 64, 1311-12).

[70] See for instance *Quodl.* q. 18 n. 3; XXVI, 230ab; *Rep. Par.* 2, d. 34, q. un., n. 3; XXIII, 170ab.

of a secondary perfection. Thus the good is also twofold and can be taken in the first and second way. Understood in the first way, good has no contrary or corresponding privation in nature; for contraries are related to one and the same thing, and therefore what is by nature incapable of being in another, does not have a contrary, nor a corresponding privation. But that which is said to be good or perfect in the sense of a primary perfection is not able to be in something else. While in so far as its being goes, it can be in another,—for an accident is said to be perfect by a primary or intrinsic perfection essential to it—nevertheless, in so far as it is a good in the sense of primary perfection, it bespeaks perfection in itself and in reference to itself. Consequently, a good in the sense of primary goodness has only the *non-good* as contradictorily opposed to it. Good in the second sense, however, namely, that which is extrinsic, has evil as its privative opposite. For evil is the privation of good according to [St. John] Damascene . . . as darkness is the privation of light. Evil therefore is opposed by way of privation to goodness taken in this second way.[71]

This first or primary goodness is identified with ontological goodness as a universal property of being, as is evident from the context. It is found wherever being is found. As such, it has no evil opposed to it, since evil is essentially a privation and as such must exist in something else. Hence its correlative good must also inhere in something, namely, in that which it perfects

---

[71] *Rep. Par.* 2, d. 34, n. 3; XXIII, 170ab: Dico quod *bonum* et *perfectum* idem sunt 6 *Metaph. Perfectum* autem dupliciter dicitur: uno modo cui nihil deest, et hoc intrinsece, et illud est perfectum perfectione essentiali intrinseca, seu perfectione prima; alio modo dicitur perfectum perfectione secunda. Sic igitur *bonum* duplex est: primo modo et secundo. Bonum primo modo non potest habere contrarium, neque privativum in natura, quia contraria nata sunt fieri circa idem; igitur quod non est natum inesse alteri, non habet contrarium, neque privative oppositum; sed bonum seu perfectum perfectione prima, inquantum primum, non est natum alteri inesse. Etsi enim quantum ad id quod est, posset alteri inesse, dicendo quod accidens aliquo modo est perfectum perfectione prima, vel intrinseca perfectione, quia est essentialis, tamen inquantum primum bonum dicit perfectionem in se et ad se. Bonum igitur bonitate prima tantum habet oppositum contradictorie, ut *non bonum;* bonum autem secundo modo, quod est extrinsecum, habet malum oppositum privative. Malum enim est privatio boni secundum Damascenum *cap.* 18. ut tenebrae luminis; isto igitur secundo modo opponitur malum privative.

and in reference to which it is said to be a good. Good as an opposite of evil, in a word, is a relative good. It is good to the extent that it perfects something other than itself. Here we have the *bonum alteri*, a relative good which implies besides the primary or intrinsic goodness (the *bonum sibi*) a relation to something outside itself. Under this aspect it can be wanted or desired by something else. A corresponding relation exists in the being perfected to the good in question as to an extrinsic goal. Hence Scotus says: "The perfection of a thing is twofold, namely, intrinsic as a form, extrinsic as an end."[72]

This aspect of appetibility is, of course, based upon the primary goodness which the being desired possesses. According to Scotus, then, it would seem that goodness as a transcendental attribute coextensive with being is certainly primarily, if not exclusively, the *bonitas in se et ad se*.

If this be the case, the question arises, Is goodness in this sense to be regarded formally as the object of the will? It is difficult to determine Scotus' position here. On the one hand he states that as the intellect tends towards its object *sub ratione veri*, so the will tends towards a thing *sub ratione boni*.[73] But this does not yet say that *bonum* is the formal object of the will any more than it says that *verum* is the adequate object of the intellect. If we understand *verum* and *bonum* in the sense of intelligibility and appetibility respectively, then there is no doubt that they are prerequisite conditions rather than the formal object. The intellect knows being and the will desires being primarily and not intelligibility or appetibility respectively. Hence Scotus might be expected to deny that *bonum* is the formal adequate object of the will, for the same reason that he denies that *verum* is the formal adequate object of the intellect. In several passages he does seem to do just this. To the objection

---

[72] *Oxon.* 4, d. 31, q. un., n. 4; XIX, 308a: Duplex autem est perfectio rei, sc. intrinseca, ut forma, extrinseca ut finis.

[73] *Oxon.* 1, d. 8, q. 4, n. 11; IX, 652b: Si ab aeterno Deus ex sui immaterialitate intelligit et vult se, et hoc sub ratione boni et veri; ergo ibi est distinctio veri et boni, rationum formalium in objectis ante omnem actum circa talia objecta.

that truth is the object of the intellect as goodness is the object of the will he replies:

> I reply that the object of the intellect and the will is one and the same, that it is being itself, which is of the essence of everything and which is predicable equally of the good and of the true, although not vice versa. For the same thing which is apprehended by the intellect is under the same *ratio* willed by the will. And they are powers of equal universality in regard to their acts.[74]

While Scotus admits that a diversity of objects may imply a difference of powers, an identity of object does not necessarily imply an identity of faculties. For instance, the sense apprehension and corresponding sense appetite may have the same formal object but they are not thereby one and the same faculty or power. For this reason Aristotle does not say that powers are the same where the object is the same, Scotus notes. He says merely that powers are distinguishable by objects as well as by acts.[75]

It should be noted that this argument is valid whether we consider goodness as *appetibility* or as the *intrinsic actualization*

---

[74] *Collatio* 36, n. 5; V, 298b: Respondeo, quod licet ex diversitate objectorum sequatur diversitas potentiarum, quia visus non potest in sonum, nec auditus in colorem, tamen ex identitate objecti non sequitur identitas potentiae; idem enim est objectum appetitivae sensitivae et apprehensivae, quia sub eadem ratione appetitur, qua apprehenditur. Unde non dixit Philosophus 9 *Metaph.* text. com. 3 quod potentiae sunt eaedem, quarum objecta sunt eadem, sed quod potentiae distinguuntur per objecta, sicut et actus.

[75] *Collatio* 35, n. 4; V, 298: Respondeo quod objectum intellectus et voluntatis est unum et idem, quod est ipsum ens, quod est de essentia cujuslibet, et aequaliter praedicatur de *vero* et *bono*, licet non e converso; nam est idem quod est apprehensum ab intellectu, et sub eadem ratione est volitum a voluntate, et sunt potentiae aequalis communitatis in actibus: sed potentiae apprehensivae, et appetitivae sensitivae est unum et idem objectum formale, et ideo licet illa formalitas, qua bonitas distinguitur a veritate, non sit veritas, est tamen ens, et e converso de veritate, ideo potest apprehendi illa bonitas ab intellectu et illa veritas potest esse volita a voluntate.—Confer also *Metaph.* 6, q. 3, n. 4; VII, 336ab; *Quodl.* q. 8, n. 14; XXV, 351b-352a; *Oxon.* prol. q. 3, n. 21; VIII, 172a.

of one's being or perfection. For the will need not desire or will a thing precisely because it has realized its own perfection, but because it is what it is, namely, a being. However, the fact that a being has realized its own perfection adds additional attractiveness to the being. It can be formally apprehended and consequently willed or desired. This seems to be what Scotus implies when he says that the good which is a *finis* for the will may be either the intrinsic or the extrinsic good.[76] Regarding Scotus' teaching on the object of the will, one should distinguish, it seems, between object in the sense of *id quod appetitur* and the *finis* or purpose *cur appetitur.* The former must be something absolute, as he points out in the Metaphysics.[77] The purpose or *finis* intended is always a good, either real or apparent, either the goodness of the thing in itself [78] or its ability to perfect the one who desires it.[79]

At any rate, Scotus seems dubious about categorically affirming that the good in general or transcendental good in particular is the proper and adequate object of the will. In answer to an argument based on the assumption that *verum* is object of the intellect as *bonum* is of the will, he says:

> The minor premise is doubtful in so far as it asserts that the good is the object of the will and false to the extent that it asserts the true is the object of the intellect, as has been said in the first book, distinction three, question three.[80]

---

[76] *Rep. Par.* 2, d. 34, q. un., n. 16; XXIII, 176a: Bonum quod est finis, est bonum utroque modo quandoque.

[77] *Metaph.* 6, q. 3, n. 4; VII, 336ab.

[78] *Oxon.* 4, d. 49, q. 5, n. 3; XXI, 173a: Actus amicitiae tendit in objectum, ut est in se bonum.

[79] *Ibid.:* Actus autem concupiscentiae tendit in illud ut est bonum mihi.

[80] *Ibid.,* q. 4, n. 11; XXI, 124a: Minor...est dubia quantum ad illam partem, quod bonum est objectum voluntatis; falsa autem quantum ad illud, quod verum est objectum intellectus, sicut dictum est in *primo lib. dist.* 3, q. 3; tam ergo major quam minor requirunt prolixiorem discussionem, quam ad praesentem quaestionem spectet; pro neutra ergo parte ponenda valet hoc medium

Binkowski in his *Wertlehre des Duns Skotus* has attempted to prove that Scotus draws a distinction between the *perfectum* and the *bonum transcendentale* on the basis of a text in the *Collationes.*[81] According to Binkowski, only the *perfectum* and not *bonum transcendentale* is to be divided into primary and secondary perfection, that is, into the intrinsic and extrinsic good referred to above.[82] Only in this sense are *bonum et perfectum idem.* The context of two pertinent passages in the *Quaestiones Quodlibetales* and the *Reportata Parisiensia* does not bear out this interpretation of Binkowski.[83] It is clear that Scotus is speaking of transcendental goodness when he speaks of primary goodness or intrinsic perfection.[84]

Of the three *conditiones concomitantes esse* unity receives the greatest attention. Truth and goodness, which play such

---

[81] *Collatio* 12, n. 5; V, 196a: Duplex est bonum, unum quod convertitur cum ente, et de illo bono tantum habet unumquodque, quantum habet de entitate. Aliud est bonum, quod est idem quod perfectum quod non incurrit aliquem defectum.

[82] *Die Wertlehre des Duns Skotus* (Berlin, Ferd. Dümmler, 1936), pp. 15-16: Auf zweierlei Art vermag man die Volkommenheit und damit diese andere Güte vom Sein auszusagen. Sie kann dem Gegenstande selbst innewohnen wie z. b. seine Form. Sie ist dann die Wesensvollkommenheit ... Diese Wesensvollkommenheit besteht so in der Unversehrheit des Gegenstandes, sie besagt die Negation der Unvollkommenheit. Damit ist also Unvollkommenheit und überhaupt jede Minderung der Vollkommenheit ausgeschlossen. Einem derart vollkommennen Gegenstande fehlt nichts. Deshalb kennt die Wesensvollkommenheit, beziehungsweise die Wesengüte innerhalb des Naturlichen kein ihr konträr oder privativ Entgegengesetztes.... Deshalb gibt est hier kein Kontrarietät oder Privation, sondern nur eine Kontradiktion, d.h. ein ganz allgemeines Nicht-vollkommen, Nicht-gut.

[83] Confer *Quodl.* q. 18, n. 3; XXVI, 230ab; *Rep. Par.* 2, d. 34, q. un., n. 3; XXIII, 170ab. Francis Lychetus in his commentary on the *Quodlibeta,* q. 18, makes the same mistake as Binkowski. Confer the Vivès edition of Scotus, XXVI, 232b. The parallel text in the *Rep. Par.* (*loc. cit.*) is much clearer on this point.

[84] For a criticism of Binkowski's position, see Fidelis Schwendinger, "Zu Binkowski's Arbeiten über die Wertlehre des D. Skotus" in *Wissenschaft und Weisheit,* IV (1937) 284-288.

an important part in the Augustinian proofs for the existence of God, lose a great deal of their importance in Scotus' metaphysics, where the task of establishing the existence of the First Being is shifted to the disjunctive transcendentals. The subject of ontologic truth receives a somewhat more detailed analysis than that of goodness, probably because of its connection with ontologism, against which Scotus directed much of his dialetics. In regard to the conception of truth as a coextensive attribute of being, a definite shift away from the exemplarism of his predecessors is noticeable. Truth is conceived as the intelligibility of being and correlated more directly with logical truth.[85]

[85] Confer Cornelio Albanese, "Intorno alla nozione della verità ontologica" in *Studi Francescani,* XII (1913-1914), 274-287.

CHAPTER VI

THE DISJUNCTIVE TRANSCENDENTALS

THE disjunctive transcendentals are placed in the third layer, immediately after the simple convertible or coextensive attributes of being, for while the single members of each disjunction are limited in their extension, the disjunction as a whole is coextensive with being as such. Thus every being without exception is either contingent or necessary, substance or accident, absolute or relative, and the like.[1]

---

[1] Scotus brings out the coextensive character of the disjunction when speaking of *necessarium vel possibile*. This disjunction, like countless others, is equivalent to a simple convertible transcendental attribute of being. "Istud disjunctum *necessarium* vel *possibile,* est passio entis circumloquens passionem convertibilem cum ente, sicut sunt multa alia illimitata in entibus." (*Oxon.* 1, d. 39, q. un., n. 13; X, 625a). Francis Lychetus in his commentary on this passage (cf. Vivès edition of Scotus' *Opera Omnia,* X, 626b) points out in what sense the disjunction is equivalent to a simple (*incomplexum*) convertible attribute such as *unum* or *verum.* "Passiones disjunctae circumloquuntur aliquam passionem convertibilem cum ente, ut *necessarium* vel *possibile;* non tamen sic intelligendo quod totum disjunctum possit esse unum incomplexum, nec unum totum incomplexum includens, imo necessario plura incomplexa includit, sed dicitur circumloqui, quia habetur loco unius incomplexi. Et in syllogismo demonstrativo potest demonstrari, ac si essent unum incomplexum, sic arguendo: omne ens est necessarium vel possibile; sed lapis est ens, ergo lapis est necessarius vel possibilis. Sed loquendo de passionibus entis incomplexis, illae secundum se sunt convertibiles cum ente, ut *unum, verum, bonum,* et hujusmodi; passio vero disjuncta tantum convertitur ratione totius disjuncti, et non ratione alicujus partis tantum, ut patet, quia haec est vera: *omne ens est necessarium vel possibile,* et haec falsa: *omne ens est necessarium,* et similiter ista: *omne ens est possibile.* Est etiam forte alia differentia inter passionem incomplexam et passionem disjunctam, quia passio incomplexa convertitur cum ente, ut est indifferens ad omnia inferiora, non addendo signum distributivum, ut ens est *unum, verum, bonum,* tamen haec non est vera absolute *ens ut ens est necessarium vel possible,* abstrahit enim ab utroque. Licet enim omne ens sit necessarium vel possibile, non tamen ens ut ens, sicut etiam haec est vera: *omne animal est rationale vel irrationale,* non tamen haec est vera: *animal ut animal, est rationale vel irrationale.*"

This coextensive character of each pair of disjunctive attributes is not the same in all instances. In some cases, for example, the two members are either formally or materially contradictories. Take the notions of dependent-independent and act-potency, for instance. A simple analysis of the concepts in question reveals that each member of the disjunction is the simple negation of the other.

With such disjunctions as prior-posterior, cause-caused, however, there is nothing to indicate *a priori* that being must be either one or the other. Prior and posterior, for one, implies a plurality and, over and above, that these several beings are not essentially equal but are essentially ordered to one another. Similarly cause-caused leaves open an alternative, namely a being that is neither cause nor caused.

The method of establishing the coextensive character of each pair will differ, therefore, in each case. Where the disjunction is between contradictories there is no need to prove the coextensive or convertible character of the pair with being. But the fact that the disjunction as such holds for all being on purely logical grounds does not yet establish that both members of the pair have any real content. Because being must be either necessary or contingent does not permit one to argue, "Therefore some being is contingent."[2] With such disjunctions as prior-or-posterior, cause-or-caused, however, there is no purely logical guarantee that because being exists, therefore being must be either prior or posterior. The convertible character of such disjunctions being purely factual, it can be established only by recourse to experience.[3] But since the notions are logical corre-

---

[2] *Oxon.* 1, d. 39, q. un., n. 13; X, 625b: Nec isto modo videtur posse concludi extremum imperfectius talis disjunctionis, non enim si perfectius est in aliquo ente, necesse est imperfectius esse in alio ente; et hoc nisi illa extrema disjuncta essent correlative, sicut causa et causatum; igitur non potest ostendi de ente per aliquid prius medium disjunctum hoc, scilicet *necessarium* vel *contingens.* Nec etiam ista pars disjuncti, quod est *contingens,* posset ostendi de aliquo supposito necessario de aliquo, et ideo videtur ista, *aliquod ens est contingens,* esse vera primo, et non demonstrabilis *propter quid.*

[3] Confer for instance the method in which Scotus establishes the disjunction: "est ergo aliquod ens prius non posterius . . . et aliquod posterius

latives, the moment one member is shown to have any real content, by that very fact the reality of the other notion is established by reason of the law of correlatives. One being, for example, cannot be prior unless another is posterior. There cannot be a " before " without an " after " and vice versa.[4]

By an ingenious manipulation of these transcendental notions with their different logical properties, the metaphysician is able to establish the existence of God from a minimum of experimental data by a relatively simple and logically valid procedure. It is just this point that endows the disjunctive transcendental attributes with their unique importance. They form the skeleton of metaphysics as a natural theology, giving it structural strength and unifying its otherwise disjointed members.

The task of the metaphysician is to analyze such attributes carefully, bringing out their logical interrelations and, welding them together into a coherent system, to use them to ascend to a knowledge of the *Primum Ens*. Scotus, it may be thought, realized this to a greater extent than did many of his predecessors and contemporaries. It explains his preference of Avicenna's Aristotelianism to that of Averroes. It was the latter who committed the unforgivable metaphysical sin, if one can speak of such a thing, when he delegated to physics the exclusive task of proving the existence of God. For much as Averroes might speak of metaphysics as a divine science, by that concession he had already sold the birthright of metaphysics and subordinated it to an inferior science. For even granting that the physicist can prove the existence of God as Prime Mover—though he is more the metaphysician than the physicist in proving that the " first mover " is really first—his proofs, based as they are upon *passiones naturales*, will always remain essentially inferior to those of the metaphysician.[5]

And in proof of his point, Scotus goes on to show why Averroes is wrong. For the metaphysician needs only to analyze the con-

---

et non prius; nullum autem quin vel prius vel posterius." *De Primo Principio*, cap. 3, in fine (M. Mueller edition, p. 63).

[4] *Metaph.* 1, q. 1, n. 49; VII, 37a.

[5] *Ibid.* 6, q. 4, n. 2; 348b.

ditions under which the attributes of being which he studies can exist. Multiplicity, dependence, composition, and the like are all transcendental attributes of being which are the object of metaphysical inquiry and analysis. Yet these attributes are all conditioned in the sense that they cannot exist unless a First Cause exists which is one, independent, simple, necessary, in act and so on. It is the task of the metaphysician to make this *resolutio plena,* as St. Bonaventure would call it, revealing the necessary implications of the attributes of being which he studies. In this way, he will not only arrive at a knowledge of God, but his knowledge will be " more perfect and more immediate " than any knowledge based upon purely physical attributes, such as motion.[6]

It is easy to understand, then, why Scotus did not bother with the famous Aristotelian argument from motion, save to point out certain weaknesses which would have to be buttressed before the argumentation would be valid.[7] Neither did he make use of the proofs based upon the notions of truth and goodness, so dear to the Augustinian tradition.[8] But he was attracted by the dis-

[6] *Oxon.* prol. q. 3, n. 21; VIII, 171ab: Dico quod Avicenna cui contradixit, bene dixit, et Commentator male. Probatur...quia per omnem conditionem effectus potest demonstrari de causa *quia est,* quam impossibile est inesse effectui nisi causa sit; sed multae passiones considerantur in Metaphysica, quas impossibile est inesse nisi ab aliqua causa prima talium entium; ergo ex illis passionibus Metaphysicis potest demonstrari aliquam primam causam istorum entium esse. Minor probatur, quia multitudo entium, dependentia et compositio et hujusmodi, quae sunt passiones Metaphysicae, ostendunt aliquod esse actu simplex independens et necesse esse. Multo etiam perfectius ostenditur primam causam esse ex passionibus causatorum consideratis in Metaphysica quam ex passionibus naturalibus, ubi ostenditur primum movens esse. Perfectior enim cognitio et immediatior est de primo ente cognoscere ipsum ut primum ens vel ut necesse esse, quam cognoscere ipsum ut primum movens.

[7] Confer Scotus' criticism of the principle " Quidquid movetur ab alio movetur." *Oxon.* 1, d. 3, q. 7, n. 27; IX, 374bff; *Oxon.* 2, d. 2, q. 10; XI, 523ff; *Oxon.* 2, d. 25, q. un., nn. 12-13; XIII, 207-209. In this connection see also Efrem Bettoni, *L'Ascesa a Dio in Duns Scoto* (Milan, Societa Editrice " Vita e Pensiero ", 1943) pp. 7-18.

[8] See for example Georg Grunwald, " Geschichte der Gottesbeweise im Mittelalter bis zum Ausgang der Hochscholastik " in *Beiträge zur Ge-*

junctive attributes of being, for he saw in what we might call the " law of the disjunctives " a relatively simple method of establishing the existence of God in a way that would satisfy the strictest demands of Aristotelian logic.

It is not easy to say definitely by whom, if by anyone, Scotus was influenced in this matter. St. Thomas, it is true, had made use of the disjunction contingent-necessary in the *tertia via,*[9] but there seems to be no indication that he was struck particularly by the disjunctive character in the sense of recognizing it as but one more instance of a more far-reaching law. In fact, the importance which has come to be attributed to this powerful argument by modern scholastics is due not so much to St. Thomas himself as to his followers. St. Thomas seems to have been more impressed by the *manifestior via,* the Aristotelian argument from motion.[10] The writings of St. Bonaventure, however, suggest a more likely source of the Scotist doctrine. In fact, in reading through Scotus' description of the law of the disjunction one is instinctively reminded of a passage in the *Quaestio disputata de mysterio Trinitatis* where St. Bonaventure has lined up ten pairs of disjunctives based upon what he calls the *decem conditiones et suppositiones per se notae.*[11] Because of its interest and importance we quote the passage in full.

> If a being that is *posterior* exists, there is also a being which is *prior,* for something posterior does not exist unless there is something prior. If then there is a universe of posterior things, it is necessary that a First Being exist. If it is necessary to posit: ' something is prior and posterior in crea-

---

schichte der Philosophie des Mittelalters, VI (Münster, 1907); Augustinus Daniels, " Quellenbeiträge und Untersuchungen zur Geschichte der Gottesbeweise im dreizehnten Jahrhundert mit besonderer Berücksichtigung des Arguments im Proslogion des hl. Anselm " in *Beiträge zur Geschichte.,* VIII (Münster, 1909); Innocenzo Gorlani, *La Conoscenza Naturale di Dio secondo la Somma Teologica di Alessandro d'Hales* (Milan, Società Editrice " Vita e Pensiero ", 1933).

[9] *Summa theol.* I, q. 2, a. 3, c.

[10] *Ibid.,* also *Summa contra Gentiles,* I, 13.

[11] A similar use of the law of the disjunction is to be found in St. Bonaventure's *Itinerarium mentis in Deum,* c. 5, n. 5; V, 309.

tures,' it is necessary that the universe of creatures implies and cries out for a first principle.

Also, if there is a being *dependent on another,* there is a being *not dependent on another,* because nothing can bring itself from non-existence to a state of existence. Therefore, a first reason for this eduction [from non-existence to existence] must necessarily exist in a first being which is not educed by another. If, then, the being dependent on another be called created being and that which is not dependent on another be called uncreated being, which is God, all the differences [i. e. differential attributes] of being imply that God exist.

Also, if a *contingent* [possible] being exists, a *necessary* being exists, since what is possible bespeaks an indifference to existence and to non-existence. But nothing can be indifferent to existence and non-existence unless it be through something which is wholly determined to existence. If then a necessary being, having no possibility at all of not existing, is none other than God, and all else has something of contingency about it, every difference of being implies that God exists.

Also, if there is *relative* being, there is *absolute* being, for what is relative can terminate only at what is absolute. But an absolute being dependent on nothing can only be something which has not received anything from another. But this is the First Being. All other beings, however, have something of dependence. Therefore, every difference of being necessarily implies that God exists.

Also, if there is being in an *imperfect* and *qualified* sense of the word, there is being in an *unqualified* sense of the word. For what is being in a qualified sense can neither exist nor be understood unless it be understood by that which is simply being; nor that which is imperfect being save by that which is perfect being, just as a privation cannot be understood save in reference to that positive entity which is had. If therefore every created being is only in part being, and only the uncreated being is simply and perfectly being, every difference of being necessarily implies the existence of God.

Also, if there is a being which exists *for the sake of another,* there is a being which exists *for its own sake,* otherwise nothing would be good. But a being which exists for its own sake can only be one than which nothing is better, which indeed is God Himself. Therefore since the universe of other being is ordered to this one, the universe of beings implies both in thought and in reality that God exists.

Also, if there is being *by participation,* there is a being that is by reason of *its essence.* For participation is spoken of only in reference to something which is possessed essentially by something, for whatever is *per accidens* can be traced back to something *per se.* Now every being other than the first being, which is God, has being by participation. He alone has being by reason of His essence; therefore, and so on. . .

Also, if a being exists *potentially,* a being that is *actual* exists; for no potency is reducible to act save by a being in act nor would there be a potency if it were not reducible to act. If, then, a being which is pure act, having nothing of potentiality, is none other than God, then every other being necessarily implies that God exists.

Also, if there is a *composite* being, there is a *simple* being, for what is composite does not have existence of itself. It is necessary therefore that it take its origin from something that is simple. But a being which is most simple, having nothing of composition whatsoever, is none other than the primary being. Therefore all other being implies that God exists.

Also, if *changeable* being exists, an *unchangeable* being exists; because according to what the Philosopher proves, motion is from a being at rest and for the sake of a being at rest. If, therefore, being that is wholly immutable is none other than the First Being, which is God, and all else is created, and in so far as created, is mutable, the existence of God is necessarily inferred from any difference of being.[12]

---

[12] *De Mysterio Trinitatis,* q. 1, a. 1; V, 46b-47b: Item ostenditur hoc ipsum secunda via sic: omne verum, quod clamat omnis creatura, est verum indubitabile; sed Deum esse clamat omnis creatura; ergo etc.—Quod autem omnis creatura clamet, Deum esse, ostenditur ex decem conditionibus et suppositionibus per se notis.

Prima est ista: si est ens posterius, est et ens prius, quia posterius non est nisi a priori: si ergo est universitas posteriorum, necesse est, esse ens primum. Si ergo necesse est ponere, aliquid esse prius et posterius in creaturis; necesse est, universitatem creaturarum inferre et clamare primum principium.

Item, si est ens ab alio, est ens non ab alio: quia nihil educit se ipsum de non esse in esse: ergo prima ratio educendi necesse est, quod sit in ente primo, quod ab alio non educitur. Si ergo ens ab alio dicitur

These arguments, which Gilson does not seem to have appreciated at their full worth in his otherwise excellent work on St.

---

ens creatum, et ens non ab alio dicitur ens increatum, quod Deus est; omnes entis differentiae inferunt, Deum esse.

Item, si est ens possibile, est ens necessarium: quia possibile dicit indifferentiam ad esse et non esse; nihil autem indifferens ad esse et non-esse potest esse nisi per aliquid, quod est omnino determinatum ad esse. Si ergo ens necessarium, nihil habens omnino de possibilitate ad non-esse non est nisi Deus, omne autem aliud habet aliquid de possibilitate, quaelibet entis differentia infert, Deum esse.

Item, si est ens respectivum, est ens absolutum: quia respectivum nunquam terminatur nisi ad absolutum; sed ens absolutum a nullo dependens non potest esse nisi quod nihil recipit aliunde; hoc autem est ens primum, omne autem aliud ens est habens aliquid de dependentia: ergo necesse est, quod quaelibet entis differentia inferat, Deum esse.

Item, si est ens diminutum sive secundum quid, est ens simpliciter: quia ens secundum quid nec esse nec intelligi potest, nisi intelligatur per ens simpliciter, nec ens diminutum nisi per ens perfectum, sicut privatio non intelligitur nisi per habitum. Si ergo omne ens creatum est ens secundum partem, solum autem ens increatum est ens simpliciter et perfectum; necesse est, quod quaelibet entis differentia inferat et concludat, Deum esse.

Item, si est ens propter aliud, est ens propter se ipsum, alioquin nihil esset bonum; sed ens propter se ipsum non est nisi ens illud, quod nihil est melius, quod quidem est ipse Deus: ergo cum universitas aliorum entium sit ordinata ad illud; universitas entium infert Deum et secundum esse et secundum intellectum.

Item, si est ens per participationem, est ens per essentiam: quia participatio non dicitur nisi respectu alicujus essentialiter habiti ab aliquo, cum omne per accidens reducatur ad per se; sed quodlibet ens aliud a primo ente, quod Deus est, habet esse per participationem, illud autem solum habet esse per essentiam: ergo etc.

Item, si est ens in potentia, est ens in actu: quia nunquam potentia est reducibilis ad actum nisi per ens in actu, nec esset potentia, nisi esset reducibilis ad actum: si ergo ens, quod est actus purus, nihil habens de possibilitate, non est nisi Deus; necesse est, quod omne aliud a primo ente inferat, Deum esse.

Item, si est ens compositum, est ens simplex: quia compositum non habet esse a se, ergo necesse est, quod a simplici recipiat originem; sed ens simplicissimum, nihil de compositione habens non est nisi ens primum: ergo omne aliud ens infert Deum.

Item, si est ens mutabile, est ens immutabile: quia, secundum quod probat Philosophus, motus est ab ente quieto et propter ens quietum: si ergo ens omnino immutabile non est nisi illud ens primum, quod Deus

Bonaventure,[13] reveal several interesting features. To begin with, they are couched in the conditional form, " if A, then B." In this form, the proposition is necessary and satisfies the first requisite for an Aristotelian demonstration, namely, that it begin with a necessary proposition.[14] Secondly, the character of the disjunction is not only recognized but is capitalized upon. The disjunctions, which in most instances are reduced to simple contradictions, are expressly referred to as *differentiae entis* or as Scotus would prefer to call them, *passiones entis* or *primae differentiae entis*. Even the argument based upon the *passio naturalis* of motion is couched in the mode of possibility. St. Bonaventure does not argue from motion to a mover but from movable to immovable or unchangeable being.

He refers to them as the " ten conditions or suppositions." This seems to refer to the conditional form in which they are formulated. The meaning of *per se notae,* however, is not so clear. Does he mean that the very analysis of one notion implies the other? This is evidently true in some instances but not in all. Or does he perhaps mean that the simple analysis of creatures reveals immediately that the antecedent of each condition is verified? The fact that he appends what may be considered a proof of the *consequentiae* to each of the disjunctions would seem to imply that they are not so self-evident, and certainly Scotus did not regard them as such, save in the case of the simple correlatives, prior and posterior. Probably the *per se notae* is to be understood in the light of his theory of illumination.

---

est, cetera autem creata, eo ipso quod creata, sunt mutabilia; necesse est, quod Deum esse inferatur a qualibet entis differentia.

Ex his igitur decem suppositionibus necessariis et manifestis infertur, quod omnes entis differentiae sive partes inferunt et clamant, Deum esse. Si ergo omne tale verum est verum indubitabile; ergo necesse est, quod Deum esse sit indubitabile verum.

[13] Gilson, *The Philosophy of St. Bonaventure,* (London, Sheed and Ward, 1940), p. 124ff.

[14] *Metaphysica,* VII, 15 (1039b 31 - 1040a 2); *Analyt. Post.* I, c. 4 (73a 23); also c. 6, 8, & 30 (passim). See also Owen Bennett, *The Nature of Demonstrative Proof According to the Principles of Aristotle and St. Thomas Aquinas* (Washington, D. C., Catholic University Press, 1943).

But what is of paramount importance, St. Bonaventure was aware of the law of the disjunctive attributes of being. It only remained for Scotus to formulate it expressly, which he did when he wrote in the Oxford Commentary:

> In the disjunctive attributes, however, while the entire disjunction cannot be demonstrated from being, nevertheless *as a universal rule by positing the less noble extreme of some being we can conclude that the more noble extreme is realized in some other being.* Thus it follows that if some being is finite, then some being is infinite. And if some being is contingent, then some being is necessary. For in such cases it is not possible for the more imperfect extreme of the disjunction to be existentially predicated of being particularly taken, unless the more perfect extreme be existentially verified of some other being upon which it depends.[15]

In formulating this law of the disjunction, as we have called it, Scotus does not say that from the existence of one member we immediately infer the existence of the more noble extreme. He uses instead the cautious term *concludi.* In many cases the reasoning is long and laborious, as for instance, in the case of finite to infinite.

The applications of this law are broad. They are not confined merely to metaphysics as a theologic, as Scotus himself was quick to see. It underlies our proofs for the reality of substance. Scotus employs it time and again in the analysis of relations with all its various ramifications.

Before indicating how Scotus attempted to prove the law of the disjunction, it will be helpful to enumerate and classify some of the disjunctives expressly mentioned by Scotus.

---

[15] *Oxon.* 1, d. 39, q. un., n. 13; X, 625ab: In passionibus autem disjunctis, licet illud totum disjunctum non possit demonstrari de ente, tamen communiter supposito illo extremo quod est minus nobile de aliquo ente, potest concludi aliud extremum quod est nobilius de aliquo ente, sicut sequitur: si aliquod ens est finitum, ergo aliquod ens est infinitum, et si aliquod ens est contingens, ergo aliquod ens est necessarium; quia in talibus non posset enti particulariter inesse imperfectius extremum, nisi alicui enti inesset perfectius extremum a quo dependeret. (Fernandez-Garcia edition, I, 1214.)

CLASSIFICATION OF THE DISJUNCTIVE TRANSCENDENTALS

To avoid any misunderstanding, it should be noted at the outset that Scotus nowhere attempts an exhaustive enumeration of the disjunctive transcendental attributes of being. In fact the almost infinite possibilities would forbid any such attempt at the very start. Scotus is content with mentioning a few instances to illustrate the difference between them and the simple convertible attributes, and noting that there are *multa alia illimitata in entibus*.[16]

The following list of disjunctives, then, has been culled from different passages of Scotus' works where he either expressly refers to the pair in question as disjunctive transcendentals or from his description of them, it is easy to see that he considered them to be such. Prior-posterior,[17] independent-dependent,[18] necessary-contingent,[19] absolute-relative,[20] infinite-finite,[21] *finiens-finitum*,[22] actual-potential, [23] simple-composed,[24] one-many,[25] cause-caused,[26] effecting-effect,[27] exceeding-exceeded,[28] substance-accident,[29] same-diverse,[30] equal-unequal.[31]

---

16 *Ibid.*

17 *De Primo Principio*, c. 3, p. 63.

18 *Oxon.* prol. q. 3, n. 21; VIII, 171b.

19 *Ibid.*, also, *Oxon.* 1, d. 39, q. un., n. 13; X, 625a; *Ibid*, d. 8, q. 3, n. 19; IX, 598ab.

20 *Oxon.* 2, d. 1, q. 4, n. 15; XI, 111a.

21 *Oxon.* 1, d. 8, q. 3, n. 19; IX, 598b.

22 *De Primo Principio*, c. 2-3, passim.

23 *Oxon.* 1, d. 8, q. 3, n. 19; IX, 598a; *Metaph.* 1, q. 1, n. 23; VII, 22a.

24 *Oxon.* prol. q. 3, n. 21; VIII, 171b.

25 *Ibid.*; also *Oxon.* 3, d. 1, q. 3, n. 3; XIV, 81ab; *Metaph.* 1, q. 1, n. 23; VII, 22a; *Quodl.* q. 6, n. 12; XXV, 249b.

26 *Oxon.* 1, d. 39, q. un., n. 13; X, 625b.

27 *De Primo Principio*, c. 2-3, passim.

28 *Ibid.*

29 *Oxon.* 1, d. 8, q. 3, n. 25; IX, 623b-[illegible]a; *Metaph.* 4, q. 1, n. 8; VII, 149b; *Ibid.*, n. 14; 154b.

30 *Meta.* 4, q. 1, n. 8; VII, 149b; *Oxon.* 1, d. 19, q. 1, n. 3; X, 170a.

31 *Oxon.* 1, d. 19, q. 1, n. 3; X, 170a.

If we compare these with the ten disjunctions mentioned by St. Bonaventure, the similarity is striking. The first four are identical, the only difference being that St. Bonaventure calls independent and dependent being *ens a se* and *ens ab alio* respectively. The *ens diminutum* and *ens simpliciter* mentioned by St. Bonaventure can be reduced to finite and infinite being, since it is clear from the context that the saint understands *diminutum* and *ens secundum quid* in the sense of imperfection and limitation. Hence it is not to be confused with the *ens diminutum* or *secundum quid* of which Scotus speaks, when referring to how an effect may be said to exist in the mind or will of its cause.[32] The disjunction *ens propter aliud* and *ens propter se*, expressing a reference to final causality is covered by Scotus' disjunction *finiens-finitum*. *Finitum* is understood here in the sense of being the object or being which is " finalized", that is, which exists for the sake of another.[33] The disjunction of St. Bonaventure based upon the notion of " participation " can be reduced to the notion of finiteness or limitation.[34] It is a distinctly Platonic notion [35] which Scotus attempts to reduce to more precise Aristotelian notions. The remaining two disjunctions, act-potency, simple-composed, are found in both. Scotus makes no mention of the disjunction *ens mutabile-ens immutabile,* since the *ens mobile* is

---

[32] See for instance *Oxon.* 1, d. 36, q. un., n. 9; X, 576b; *Oxon.* 1, d. 39, n. 21, X, 637ab.

[33] *De Primo Principio,* c. 1, p. 8: Primum membrum secundae divisionis quod est causa, famose subdividitur in quatuor causas satis notas; finalem et efficientem, materialem et formalem. Et posterius sibi oppositum dividitur in quatuor correspondentia, scilicet in ordinatum ad finem, quod, ut breviter loquar, dicatur finitum; et in effectum; et in causatum ex materia, quod dicatur materiatum; et in causatum per formam, quod dicatur formatum.

[34] *Quodl.* q. 5, n. 26; XXV, 229b-230a: Omne finitum cum sit minus illa entitate infinita, conformiter potest dici pars, licet non sit secundum aliquam proportionem determinatam, quia exceditur in infinitum, et hoc modo omne aliud ens ab ente infinito dicitur ens per participationem, quia capit partem illius entitatis, quae est ibi totaliter et perfecte. Hoc volo habere, quod omne finitum, cum sit minus infinito, est pars.

[35] Hirschberger, " Omne ens est bonum " in *Phil. Jahrbuch,* LIII (1940) 297-298.

not properly speaking the object of metaphysics but of physics. The remaining disjunctions either fall under some of the previous as subdivisions, or they do not lead us directly to God, for instance, one-many, substance-accident, same-different.

As Scotus has not attempted a definite enumeration of these transcendentals, neither has he attempted to classify them. For his purposes it seemed enough to indicate that an essential order exists between the two members of every disjunction, and to give the principles governing such an order. It is in this latter analysis, as will be seen later, that the fundamental reason for the law of the disjunction is to be found.

Allusion has been made already to the two distinct types found among the notions mentioned by Scotus. The one group being primarily correlatives, the other being contradictories. Using this distinction as a tentative basis of classification, we can proceed to discuss the nature of the principal disjunctives mentioned by Scotus.

I. *The Correlative Disjunctions*

The correlative disjunctions mentioned by Scotus are prior-posterior, cause-caused, exceeding-exceeded. The second of these three is subdivided into effecting-effected, *finiens-finitum.*

1. *Prior or Posterior*

Prior and posterior are correlatives. As such they presuppose the notion of relation in general, hence the notions of plurality and of inequality. Relations of inequality constitute an order. Hence to establish that *Omne ens est prius vel posterius* is equivalent to verifying the proposition *Omne ens est ordinatum.*[36]

Scotus understands *ordinatum* or ordered not in the strict technical sense in which what is prior or first is said to be *extra ordinem* and only what is posterior is said to be ordered. But he takes it in the most general sense as applying to the mutual relation between what is prior and what is posterior.[37]

---

[36] *De Primo Principio,* c. 1, p. 2; c. 3, p. 63.

[37] *Quodl.* q. 19, n. 2; XXVI, 260a: Ordo autem est posterius ad prius; *De Primo,* c. 1, pp. 2-3: Accipio autem ordinem essentialem, non stricte—ut quidam loquuntur, dicentes posterius ordinari, sed prius vel primum esse supra ordinem—sed communiter, prout ordo est relatio aequiparentiae,

The order of priority and posteriority as coextensive with being is understood to be an essential order,[38] that is, an order based on the very essence or nature of the beings in question.[39]

Two beings are by their essence unequal either because the one owes its being or its existence to the other or because one has simply more being or more perfection than another without, however, depending necessarily on the other. In the first case the order will be one of dependence, in the other it will be one of eminence.[40]

2. *Exceeding or exceeded*

In the order of eminence, that which is prior is the being which exceeds the other in perfection and nobility (the *excedens*) while that which is exceeded (the *excessum*) is the posterior.[41] The idea of more perfect and more noble introduces a notion that requires further clarification. For perfection and nobility are taken in reference to something absolute and consequently seem to imply a previous knowledge of what is all perfect. This problem will recur in connection with the pure perfections. For the present we may understand it simply as a difference in the amount or quality of being.

3. *Cause or caused*

In the order of dependence, that which is prior is that on which something depends (*id a quo dependet*) while the posterior is the

---

dicta de priori respectu posterioris, et e converso, prout scilicet ordinatum sufficienter dividitur per prius et posterius. Sic igitur quandoque de ordine, quandoque de prioritate vel posterioritate fiet sermo.

[38] See the context of the first and third chapters of the *De Primo Principio*.

[39] *Quodl.* q. 19, n. 5; XXVI, 266a: Ordo essentialis est per se inter essentias.

[40] *De Primo,* c. 1, p. 3: Dico ergo primo, quod ordo essentialis videtur primaria divisione dividi, sicut aequivocum in aequivocata, in ordinem eminentiae, et in ordinem dependentiae.

[41] *Ibid.:* Primo modo prius dicitur eminens, et posterius, quod est excessum. Ut breviter dicatur: quidquid est perfectius et nobilius secundum essentiam, est sic prius.

dependent being.[42] Scotus reduces the relationship of dependence to that of causality, but with this significant observation. The being which can cause does not necessarily require the effect, but the effect necessarily requires the cause.[43] In a word Scotus stresses the independence of the cause in regard to its corresponding effect. The fact that all created causes are themselves caused introduces an essential element of contingency into secondary causality. While we may distinguish between free and necessary causes in created being, necessary is understood only in a relative sense. For this reason Scotus points out elsewhere that it is a contradiction that any cause should cause necessarily in the unqualified sense of the term, a position he holds in opposition to Aristotle.[44]

The priority which the cause has over the thing caused is one of nature and not necessarily of duration. In fact, in the case

---

[42] *De Primo,* c. 1, pp. 3-4.

[43] *Ibid.:* Prius dicitur, a quo aliquid dependet, et posterius, quod dependet. Hujus prioris hanc intelligo rationem, quam et Aristoteles 5 Metaphysicae testimonio Platonis ostendit: prius secundum naturam et essentiam est, quod contingit esse sine posteriori, non e converso. Quod ita intelligo, quod, licet prius necessario causet posterius et ideo sine ipso esse non possit, hoc tamen non est, quia ad esse suum egeat posteriori, sed e converso; quia si ponatur posterius non esse, nihilominus prius erit sine inclusione contradictionis; non sic e converso, quia posterius eget priore, quam indigentiam possumus dependentiam appellare, ut dicamus, omne posterius essentialiter a priore necessario dependere, non e converso, licet quandoque necessario posterius consequatur istud.

[44] *Oxon.* 2, d. 1, q. 3, n. 12; XI, 77b-78a: Simpliciter enim necessario causare includit contradictionem. *Ibid.* n. 15; 79b: Ad illud quod additur de Philosophis, potest dici, quod multas contradictiones latentes concesserunt, sicut negaverunt communiter esse aliquod primum principium contingenter causans, sed dixerunt primam causam esse necessario causantem, et tamen dixerunt contingentiam esse in entibus et aliqua contingenter fieri; sed contradictionem includit aliquid contingenter fieri in entibus, et primam causam necessario causare. *Quodl.* q. 7, n. 43; XXV, 317a: Quando ergo probatur secundum intentionem Aristotelis, Angelum non esse causatum, quia secundum ipsum est formaliter necessarium, dico quod ipse non posuit ista inter se repugnare causatum et formaliter necessarium, cum dicat 2 Metaph. *Sempiternorum principia semper esse verissima necesse est, quia sunt aliis causa veritatis.* Sempiterna ergo, quae ipse posuit formaliter concessit principia habere.

of an essential order of dependence the simultaneous existence of the cause and the thing caused is required.[45] In regard to the opposition between cause and caused, note that it is one of contrariety rather than of contradiction. Consequently, in an essential order of dependence it is repugnant that one and the same thing could depend upon itself.[46] This presupposes that an essential order exists between the two. It is not repugnant for instance that the mover and the thing moved (*motum*) be the same, for there is only an accidental order between the *ens mobile* and its actual state of motion in so far as it is the recipient of the motion.[47] This does not say that motion in so far as it is an accidental entity does not essentially depend on its respective cause. This, however, is another question.

Scotus divided the causes according to the classical division into formal, material, final and efficient. An essential order exists in each of these types of causality, but the first two involve imperfection in both of their members.[48] Consequently, they cannot be coextensive in disjunction with all being. Efficient and final causation involve no imperfection on the part of the cause.[49] The efficient cause is defined as one which gives exist-

---

[45] *Oxon.* 1, d. 2, q. 3, n. 4; VIII, 496a: A nullo aliquid dependet essentialiter, quo non existente nihilominus esset; *Oxon.* 1, d. 28, q. 3, n. 10; X, 424a: Causa inquantum causa, prior sit natura causato inquantum causatum, et tamen causa inquantum causa est simul causato simultate requisita ad correlativa. *Oxon.* 2, d. 1, q. 3, n. 17; XI, 83a.

[46] *Oxon.* 1, d. 3, q. 7, n. 30; IX, 376b: Causa et causatum in eadem natura sive supposito repugnant, quia si non, tunc idem dependeret a se.

[47] *Ibid.;* Movens autem et motum nec in eadem natura, nec in eodem supposito repugnant, quia hic non ponitur dependentia essentialis, qualem ponunt relationes causae et causati.

[48] *Oxon.* 1, d. 5, q. 2, n. 9; IX, 472b: Causalitas causae materialis non dicit perfectionem simpliciter. *Oxon.* 1, d. 8, q. 1, n. 2; IX, 565ab: Causalitas materiae et forma includit imperfectionem, quia rationem partis; causalitas efficientis et finis nullam imperfectionem includit, sed perfectionem; omne autem imperfectum reducitur ad perfectum sicut ad prius se essentialiter, ergo causalitas materiae et formae non est simpliciter prima.

[49] *Oxon.* 1, d. 8, q. 1, n. 2; IX, 565ab; *Ibid.* 2, d. 1, q. 2, n. 7; XI, 63b: Dico quod causa potest primo et immediate aliquem effectum novum producere, absque omni novitate in ipsa.

ence to its effect.[50] The final cause exercises its causality by moving the efficient cause to give the effect its respective being.[51]

The three sets of correlatives are interrelated and can be reduced to some form of priority and posteriority. But it is not evident from a simple analysis of the terms that either must exist, nor that the disjunction is coextensive with all being.

In the *De Primo Principio*, however, Scotus has demonstrated both these points.[52] It is beyond the scope of this work to discuss in more than brief outline the procedure employed in this *tractatus aureus* unequaled in scholastic literature for its orderly and consistent reasoning.

By a cleverly concatenated, yet relatively simple, series of proofs, Scotus analyzes the various essential orders and their mutual implications. He shows that nothing can be essentially ordered to itself but only to something distinct from itself. That further, a circle or an infinite regress in the order of essential causes is impossible. Consequently, in every essential order, there must be a first being. An analysis of the interrelation between the various essential orders further reveals that if an order of efficient causality exists, an order of final causality and of eminence must exist, and furthermore, that the first in each order coincides in one and the same being.

With this background in mind Scotus proceeds with the barest of experimental knowledge, namely, *aliquid est effectibile*, to the conclusion that *omne ens est ordinatum* and that

> there is consequently some being which is prior and not posterior and hence first, and some which are posterior and not prior, but no being which is neither prior nor posterior. Thou [O God,] art the only First One. All else is posterior

[50] *Oxon.* 3, d. 6; q. 1, n. 3; XIV, 309a: Causalitas autem causae efficientis et conservantis non terminabatur nisi ad aliquod *existere* non increatum, quia nihil efficit se. Confer also *Oxon.* 4, d. 12, q. 1, n. 26; XVII, 559a.

[51] *Oxon.* 1, d. 1, q. 5, n. 6; VIII, 385b: Causalitas causae finalis est movere efficiens ad agendum. *Ibid.* d. 2, q. 2, n. 16; 419b: Causa finalis non causat nisi quia metaphorice movet ipsum efficiens ad efficiendum.

[52] *De Primo Principio*, c. 1-3.

to Thee, and this in the threefold order [namely of eminence, efficient and final causality].[53]

II. *The Disjunctive Transcendentals Contradictorily Opposed*

These differential notions are opposed in such a way that any given being must logically fall into either of the two classes. To deny that a certain being falls into one group is necessarily to affirm that it is in the other group. In other words, the two members in reference to being are not merely contrarily but contradictorily opposed. The problem may arise whether or not these differences add new positive entity to the entity represented by the indifferent concept of being. There is no doubt that the differential concept adds to the concept of being, but this does not necessarily imply that the objective basis of the differential concept need be anything more than a real privation. A special difficulty seems to arise in the case of contingency.

1. *Every being is either actual or potential*

The disjunctive character of potential and actual being[54] is expressed by the axiom *Actus et potentia dividunt ens et quodcumque entis genus.* Scotus has devoted particular attention to an analysis of act and potency as primary differences of being because of its implications in the faculty theory of the soul. He eliminates three meanings commonly given to either act or potency which do not apply here.

First of all, act and potency do not refer to mutually complementary active and passive principles that constitute a composite

---

[53] *Ibid.* c. 3, p. 63: Vere, Domine: omnia in sapientia ordinata fecisti, ut cuilibet intellectui rationabile videatur, quod omne ens est ordinatum, unde absurdum fuit philosophantibus ordinem ab aliqua amovere. Ex hac autem universali—omne ens est ordinatum—sequitur, quod non omne ens est posterius et non omne prius; quia utroque modo vel idem ad se ordinaretur, vel circulus in ordine poneretur; est ergo aliquod ens prius non posterius, et ita primum; et aliquod posterius et non prius; nullum autem quin vel prius vel posterius. Tu es unicum Primum; et omne aliud a te posterius est te, sicut in triplici ordine, ut potui, declaravi.

[54] *Metaph.* 4, q. 2, n. 10; VII, 163a: Hoc disjunctum convertit sicut potentia vel actus cum ente.

being,[55] such as, for instance, matter and form in regard to a physical composite [56] or genus and specific difference in regard to a metaphysical composite.[57] Act and potency in this sense are not mutually incompatible and can be realized in one and the same physical being. Consequently, Scotus finds no contradiction in supposing that matter is actual in the sense of having its own proper entity and existence and, at the same time, is potential in the sense that it is the receptacle of a form and together with the form constitutes an essential unity.[58]

---

[55] *Oxon.* 1, d. 26, q. un., n. 35; X, 324b: Dico quod est ibi fallacia aequivocationis. Uno modo enim actus est differentia opposita potentiae, et hoc modo dividit omne ens. Alio modo actus cum potentia constituit aliquod totum, sicut loquitur Philosophus 8 *Metaph.* de potentia et actu, quod non est verum de potentia opposita actui, quia illa non manet cum actu.

[56] *Oxon.* 2, d. 12, q. 2, n. 7; XII, 603ab: Dico quod si accipiatur actus, prout distinguitur contra potentiam secundum quae, scilicet actum et potentiam, totum ens dividitur, sic actus non convertitur cum forma. Secundum hoc enim, omne quod est extra causam suam, est in actu, et secundum hoc etiam privationes dicuntur esse actu; unde caecitas dicitur esse actualiter in oculo carente visu, sicut dicitur in *quarto et septimo Metaphys.* Si autem loquaris de actu, secundum quod loquitur Philosophus 7 *Metaphys.* scilicet secundum quod est actus receptus, et actuans et distinguens, sic distinguitur contra receptivum, et materia est receptivum illo modo, et non est actus.

[57] *Oxon.* 2, d. 2, q. 10, n. 12; XI, 540ab: Ad primum argumentum principale dictum est d. 3, *primi,* quomodo potest aliquid agere in se, et responsum est ibi ad illud argumentum. Quod autem additur pro confirmatione, quod quaedam dividentia ens sunt incompossibilia in quocumque, igitur et ista, concedo de istis ut sunt opposita. Opposita autem sunt, prout dicunt modos cujuslibet entis, prout scilicet idem est ens in potentia antequam actu sit, et ens in actu, quando jam est. Et isto modo nulli eidem conveniunt, nec formaliter, nec etiam denominative, quod scilicet idem dicatur ens denominatum simul ab aliquo in aliquo actu, et ab eodem in potentia. Ut tamen actus accipitur pro principio activo, et potentia pro principio passivo, quae cadunt infra essentiam cujuslibet definibilis vel definiti, sic non sunt opposita, nec sic dividunt ens, nec repugnat alicui eidem.

[58] *Oxon.* 4, d. 11, q. 3, n. 11, XVII, 358ab: Et quando arguitur contra primum, quia tunc materia esset sine forma, et sic in actu, et non in actu, aequivocatio est de actu, quia uno modo actus est differentia entis opposita

Secondly, potency is not taken in the sense of a power to act (active potency) nor necessarily as a subjective capability of receiving some further perfection (passive potency). This is clear from the objections Scotus finds to the principle, *Actus et potentia sunt in eodem genere*, when used to prove the real distinction between soul and faculties.[59]

Thirdly, potency is not to be taken in the sense of mere logical possibility or compatibility of notes.[60] Such a potentiality not only can, but must, coexist with whatever is actual. It enters into the very definition of being as has been seen. It is of such possibility that the axiom holds: *ab esse ad posse valet illatio.*

Act and potency, or better, actual and potential, in so far as they represent primary differences of being have reference to *actual existence.* Consequently, only a being which is dependent for its existence on another can be said to be potential. A being (namely, a quiddity to which existence is compatible) is said to be a potential being when it lacks actual existence. It " exists " only in an improper sense of the word (*secundum quid*), in so far as it is considered an *ens diminutum* virtually in its causes.[61] That which was potential is said to " become " actual when it " receives " actual existence, that is, when it has *esse extra causam suam.*[62]

---

potentiae, prout dividitur omne ens, quod est scilicet in actu, vel in potentia. Alio modo actus dicit habitudinem illam, quam dicit forma ad informabile, et ad totum cujus est. Et eodem modo aequivocatur de potentia, quia ut opponitur actui primo modo, dicit ens diminutum, cui scilicet non repugnat esse extra causam suam; ens autem in actu oppositum isti potentiae, est ens completum in suo *esse* extra causam suam, quodcumque sit illud. Alio modo potentia dicit principium receptivum actus secundo modo dicti, sicut materia dicitur potentia et forma actus.

[59] *Oxon.* 2, d. 16, q. un., n. 5; XIII, 25b ff.

[60] *Oxon.* 1, d. 39; q. un., n. 16; X, 629a: Potentia logica, quae est non repugnantia terminorum...; *Ibid.* d. 2, q. 7, n. 10; VIII, 529b: Possibile Logicum est modus compositionis formatae ab intellectu, illius quidem cujus termini non includunt contradictionem, et ita hoc modo possibilis est haec propositio: *Deum posse produci* et *Deum esse Deum.*

[61] *Oxon.* 4, d. 11, q. 3, n. 11; XVII, 358b: Dicit ens diminutum, cui scilicet non repugnat esse extra causam suam.

[62] See note 56.

Act and potency, then, as primary differences of being, are clearly existential modes, and differ from act and potency as essential or constitutive elements of things. What actually exists is called *ens in actu;* what does not actually exist but can exist, *ens in potentia.* The latter still verifies the notion of being, namely, " that to which existence is not repugnant." [63]

This analysis reveals the contradictory nature of these primary differences. It becomes clear that since actual existence does not enter into the formal notion of being as a quiddity, it cannot change the quiddity or kind of being. Hence, *Actus et potentia sunt in eodem genere,* and, as Scotus points out further, not only is the actual being and its corresponding potential being in the same genus but they are numerically identical.[64] In other words, the particular or individual object which is now said to be actual, was prior to its actual existence, potential.[65]

Potency clearly involves imperfection in this sense. Being less perfect than its corresponding actual being, it will always be related to the latter by at least an essential order of eminence.

2. *Every being is either dependent or independent*

Mention has been made of the notion of dependence in connection with the correlatives cause-caused. A dependent being is defined as one which is conditioned by a second being in such a way that the former cannot exist without the latter, but the latter can exist without the former.[66] The need or *indigentia* which the conditioned being has for that by which it is condi-

---

[63] *Oxon.* 4, d. 8, q. 1, n. 2; XVII, 7b: Ens, hoc est cui non repugnat esse.

[64] *Oxon.* 1, d. 7, q. un., n. 21; IX, 545a: Ad quartum de potentia et actu, dico quod aequivocatur de potentia, vera est enim major [sc. potentia est ejusdem generis cum actu] ut potentia est differentia entis condividens ens contra actum, quia sic ens in communi non tantum dividitur per actum et potentiam, sed etiam quodcumque genus entis, et quaecumque species et individuum, quia sic eadem albedo est in potentia et postea in actu. Et hoc modo ad idem genus pertinent actus et potentia.

[65] *Oxon.* 2, d. 16, q. un., n. 5; XIII, 25b: Alio modo accipitur [sc. potentia] ut dividitur contra actum, et hoc modo potentia et actus non sunt tantum ejusdem speciei, sed etiam ejusdem numeri, ut dicitur 9 *Metaph.* Illud enim individuum, quod nunc est in actu, illud idem fuit in potentia.

[66] Confer note 43.

tioned is called dependence. Essential dependency can be reduced to one of the four orders of causality.[67]

Dependency, then, requires as its correlative the notion of a being upon which something is dependent. This second being must have by definition a relative independence, that is, it conditions the dependent being and not vice versa. It does not follow immediately that it is totally independent. The latter simply negates all posteriority in the order of essential conditions. But it is possible to show that the dependence can only be accounted for on the condition that an absolutely independent being exist. Scotus does this by showing that a first condition must exist in every essential order of dependence. He next proves that the first condition in the order of material and formal causality is itself conditioned in reference to an efficient and final cause. Finally, he shows that the first in these two remaining orders must coincide.[68]

Furthermore, since dependence implies imperfection and limitation in the being which is dependent,[69] it is possible to show that whatever exists as the result of causality must be exceeded in perfection by at least one other being. But this is equivalent to proving that the highest in the order of perfection must be totally independent and unconditioned.[70] Where independent is

---

[67] *De Primo,* c. 1, p. 8; *Oxon.* 4, d. 49, q. 2, n. 30; XXI, 54a: Nihil videtur dependere essentialiter ab eo quod non est aliqua causa ejus, loquendo de actu primo quocumque. *Quodl.* q. 7, n. 40; XXV, 315a: Nihil est dependens ab aliquo in essendo, a quo non habet in aliquo genere causae esse. *Oxon.* 1, d. 8, q. 5, n. 7; IX, 742a.

[68] *De Primo Principio,* passim. *Oxon.* 1, d. 12, q. 1, n. 14; IX, 866b: Omne dependens dependet ad aliquid omnino et simpliciter independens, (nunquam enim dependentia alicujus sufficienter terminatur, nisi ad aliquid omnino independens).

[69] *Quodl.* q. 19, n. 4; XXVI, 265a: Dependentia vel est formaliter imperfectio vel omnino habet necessario imperfectionem annexam.

[70] See *De Primo Principio* where Scotus shows that material and formal causality imply efficient and formal though not vice versa, and that wherever we have an effect of efficient causality the effect is also conditioned by a final cause. It is therefore *finitum* in the sense of being "finalized". Hence his conclusion: "Quod omne finitum est excessum" (ch. 2, concl. 16, p. 33) implies that whatever is the result of any kind of causality is exceeded in perfection by some being.

taken in an absolute sense, it is possible to infer that an order of eminence exists between the two members of the disjunction, even though it does not follow immediately that everything which is dependent is less perfect than that on which it depends, nor that everything which is less perfect is essentially dependent upon that which is more perfect.[71]

3. *Every being is either necessary or contingent*

To understand the contradictory opposition involved in the notions of a necessary and contingent being, it is necessary to correlate contingency with the two classical logical modes of possibility and impossibility. Given a metaphysical interpretation, the logical mode of possibilty would seem to express compatibility of any given being or quiddity with existence; impossibility would express an incompatibility with existence based upon an inner incompatibility of notes, for example, a triangular sphere. The former would be equivalent to the notion of being (*hoc est, cui non repugnant esse*), the latter would be a nothing (*nihil simpliciter*), having only a *quid nominis*.

Contingent and necessary as primary differences of beings, therefore, are further differentiations of the logical possibility of existence.[72] This is probably why Scotus speaks of both contingency and necessity as positive modes,[73] for both further

---

[71] *Ibid.* c. 2, conclusions 13 and 14, pp. 30-31.

[72] *Oxon.* 1, d. 8, q. 5, n. 7, IX, 742b: Necessitas enim est conditio existentiae. *Ibid.* d. 18, q. un., n. 4; X, 144b: Creatura vero in nullo *esse,* nec intra nec extra est necesse esse, neque in *esse* existentiae nec in *esse* essentiae, scilicet diminuto sive intellecto. Ratio hujus est, quia necessitas est conditio existentis, et tale *esse* [sc. diminutum] abstrahit ab existentia.—This passage is interesting because it indicates that necessity (and consequently contingency) are modes of existence. Creatures, in so far as they exist in the divine intellect as archetypal ideas, are neither formally necessary nor contingent, even though the act of the divine intellect by which they are produced in *esse intellecto* is necessary. Necessity and contingency, then, apply only to true existence (*esse simpliciter*) and not properly speaking to the *esse secundum quid,* i.e. the *esse intellectum.*

[73] *Oxon.* 1, d. 39, q. un., n. 35; X, 656b: Dico quod contingentia non est tantum privatio vel defectus entitatis, sicut est deformitas in actu secundo, qui est peccatum. Imo contingentia est modus positivus entitatis, sicut necessitas est alius modus; et omne positivum, quod est in effectu,

specify the nature of the union of a given quiddity with existence. Where the being is necessary, this union is such that it could simply not be otherwise. There is no condition under which it would be true that this thing did not exist. A contingent being on the contrary is anything "whose opposite could have been when this came to be."[74] "Necessity, however, deprives completely all possibility that the opposite of what is should be."[75]

From the standpoint of causality, Scotus refers to a contingent being as one which can have *esse post non esse*.[76] The meaning of this is best understood by considering the source of contingency, namely, a First Cause which causes its effects contingently—the divine will of God.[77] Characteristic of this cause is the peculiar way in which it tends towards its object to give it existence, namely "it contingently tends to this object in such a way that in the same instance it could tend to an opposite object."[78] This does not imply that God could will that a given object exist and not exist at the same time, but that He has simultaneously the power to will either the existence or non-existence of the thing. *Simul habet potentiam ad opposita, sed non habet potentiam ad opposita simul.*[79]

---

principalius est a causa priore, et ideo non sicut deformitas est ipsius actus a causa secunda, et non a causa prima, ita est contingentia, imo contingentia per prius est a causa prima quam secunda, propter quod nullum causatum esset formaliter contingens, nisi a causa prima contingenter causaret sicut ostensum est. (The positive character of the mode of necessity also follows from its character as a pure perfection.) *Oxon.* 1, d. 2, q. 7, n. 6, VIII, 520a: Necessitas est simpliciter perfectionis.

[74] *De Primo Principio*, c. 4, p. 74: Dico hic contingens... cujus oppositum posset fieri quando istud fit.

[75] *Oxon.* 1, d. 39, q. un., n. 31; X, 653b: Necessitas autem simpliciter privat absolute possibilitatem hujus oppositi.

[76] *De Primo Principio*, c. 3, p. 37: Quia aliqua est contingens, igitur possibilis esse post non esse.

[77] *Oxon.* 1, d. 39, n. 35; X, 656b. (Confer note 73.)

[78] *Ibid.*, n. 22; 637b.

[79] This is true of free will in general. As Scotus nicely puts it in regard to our will: "Simul habeo potentiam ad opposita, sed non ad opposita simul." (*Ibid.* n. 20; 632a).

Though existence is so conditioned that it must be either necessary or contingent, the positive character of contingency makes it necessary to qualify the statement that necessity is a more perfect condition than contingency. It is more perfect, says Scotus, only where it is compatible with the entity in question. It is incompatible with the formal notion of cause as such. Therefore, necessity is incompatible with the divine will in its creative activity *ad extra,* not, however, in its internal operations.[80]

4. *Every being is either substance or accident*

Scotus does not discuss substance and accident precisely as transcendental disjunctives except to indicate that substance can be understood as a category " and in this way includes limitation; but substance is accepted there [i. e. when applied to God] for a being in itself and not in another." [81] In the latter case, substance abstracts from what is finite and infinite and hence is " more universal than the concept of substance as a genus." [82] Substance is not simply to be defined as *ens non in alio* but rather as a being *cui convenit non inhaerere, vel cui repugnat inhaerere* and an accident as *natura cui convenit inhaerere.*[83] In this form, the contradictory character of the two notions is apparent.[84] Substantiality and accidentality refer to existence

---

[80] *Oxon.* 1, d. 8, q. 5, n. 25; IX, 765a: Dico quod necessarium est conditio perfectior in omni entitate quam possibile, cui conditio necessitatis est compossibilis. Non est autem perfectio in illa entitate, cui non est compossibilis, quia contradictio non ponit aliquam perfectionem. *Ibid.*, n. 12; 749b-750a: Necessitas est perfectior conditio ubi est possibilis; est autem incompossibilis rationi causae ut causa, quia sic loquimur, et non de eo quod est causa.

[81] *Oxon.* 1, d. 8, q. 3, n. 25; IX, 623b-624a: Dico quod si substantiam abstrahas a creata et increata, non accipitur ibi *substantia* ut est conceptus generis generalissimi, quia increata repugnat substantia hoc modo quia substantia hoc modo includit limitationem; sed accipitur ibi *substantia* pro ente in se, et non ente in alio, cujus conceptus prior est, et communior conceptu substantiae, ut est genus, sicut patuit per Avicennam.

[82] *Ibid.*, 624a.

[83] *Quodl.* q. 3, n. 19; XXV, 140b-141a.

[84] See also *Metaphy.* 4, q. 2, n. 11; VII, 163b.

rather than to the quiddity itself. An essential order, it should be noted, obtains between accident and substance.[85]

5. *Every being is either finite or infinite*

Finite and infinite express what might be called the amount or the "quantity" of perfection rather than the kind or quality of perfection. "Quantity" here is obviously not to be understood in the sense of predicamental quantity, but rather in the sense of representing an answer to the question "how much?" Likewise quality is understood in the sense of an answer to the question "what kind of being?"[86] This quantity is more frequently referred to as *quantitas virtualis* or *magnitudo*.[87]

---

[85] *Ibid.*, n. 12; 164a: Est ordo essentialis quantitatis ad substantiam, et in comparatione ad substantiam, dicitur quantitas prior aliis accidentibus, quia immediatius inest substantiae, quam aliquod aliud accidens; ordo autem aliquorum accipitur secundum id quod est in ipsis essentiale sive principale, non secundum aliquid accidentale ipsis.

[86] *Quodl.*, q. 5, n. 26; XXV, 230ab: Finitum tamen et infinitum non dividunt ens, nisi ens quantum, quia sicut secundum Philosophum *primo Physicorum,* finitum et infinitum quantitati congruunt, quod est verum de finito et infinito, et quantitate proprie acceptis, ita etiam extensive loquendo, finitum et infinitum, ut sunt passiones entis, conveniunt praecise enti quanto in se habenti quantitatem aliquam perfectionalem; talis autem quantitas non convenit entitati, nisi quae potest esse partialis vel totalis inter essentias.

[87] *Ibid.*, 230b: Quantitas virtualiter non convenit nisi entitati quidditativae, ut scilicet distinguitur ab entitate hypostatica, et per consequens nec finitas, nec infinitas. *Oxon.* 1, d. 19; q. 1, n. 3; X, 169b-170a: Et secundum hoc sicut potest considerari quodcumque ens, ita etiam super ipsum potest fundari triplex relatio communiter sumpta, quia identitas super quodcumque ens inquantum est *quid,* et aequalitas vel inaequalitas super quodcumque ens, inquantum habet magnitudinem aliquam perfectionis quae dicitur quantitas virtutis, de qua dicit Augustinus 6, de *Trinit.* c. 7, quod in his 'Quae non magna mole sunt, idem est medius esse quod majus esse.' *Ibid.*, n. 6, 171b: Dico quod non est ibi [sc. in Deo] quantitas molis, sed virtutis; et si nomen *quantitatis* appropriatur ad magnitudinem molis, magnitudo autem non appropriatur ad magnitudinem molis; tunc potest proprie concedi, quod est magnitudo ibi sine quantitate, et ita magnitudo vere est fundamentum aequalitatis transcendentis, quia hoc modo omne ens est magnum vel parvum aequale vel inaequale, licet magnitudo ista non sit fundamentum aequalitatis, prout est passio quantitatis, quae est genus.

"Every being is something in itself and has in itself some determinate grade among beings." [88] The magnitude of its perfection will be either limited or without limit. In the first case it is said to be finite, in the second infinite.[89] While Scotus' discussion of relations and personality in the Trinity has complicated to some extent the clarity of the disjunction of infinite and finite as a formal contradiction, nevertheless he admits that for all quidditative being, and for everything that is *a being*, the contradictory character of infinite and finite holds, so that whatever is not a limited or finite being is infinite.[90]

Infinite and finite are singled out particularly by Scotus as examples of intrinsic modes.[91] They are essential modes rather than existential,[92] and are intimately bound up with the actually existing perfection. For this reason, every perfection in God may be said to be formally infinite.[93]

---

[88] *Ibid.*, n. 3; 169b: Dico quod quodcumque ens est in se quid et habet in se aliquem gradum determinatum in entibus.

[89] See *Quodl.* q. 5, n. 4; XXV, 199b-200b; *Ibid.* n. 26; 229b: Infinitas in entitate dicit totalitatem in entitate, et per oppositum suo modo finitas dicit partialitatem entitatis. Omne enim finitum ut tale, minus est infinito ut tali.

[90] *Oxon.* 1, d. 8, q. 3, n. 16; IX, 596a: Quidquid dicitur communiter de Deo et creatura, est indifferens ad finitum et infinitum (loquendo de essentialibus), vel saltem ad finitum et non finitum (loquendo de quibuscumque), quia relatio divina, nec est finita, nec infinita; nullum autem genus potest esse indifferens ad finitum, et infinitum; ergo, etc.... quidquid est in Deo perfectio essentialis est formaliter infinitum, in creatura vero finitum.—Confer also *Quodl.* q. 5: "Utrum relatio originis sit formaliter infinita" (*tota quaestio*); XXV, 198-230.

[91] *Oxon.* 1, d. 8, q. 3, n. 27-29; IX, 626b-629b; *Ibid.*, d. 19, q. 1, n. 5; X, 171a: Istud autem magnitudo non est aliud attributum a perfectione tali, quia dicit modum intrinsecum illius sicut saepe supra dictum est. *Rep. Par.* prol. q. 1, n. 44; XXII, 30a.

[92] *Quodl.* q. 5, n. 4; XXV, 200b: Infinitas est magis modus intrinsecus essentiae quam aliquod attributum. *Oxon.* 1, d. 8, q. 3, n. 28: IX, 629a: Infinitas esse modus entitatis per essentiam.

[93] *Oxon.* 1, d. 8, q. 3, n. 25; IX, 624a: Omnis autem realitas in Deo est infinita formaliter. (Note: *Realitas* is here understood of any formal perfection. The *entitas hypostatica* is not a perfection in the proper sense

It is obvious that an essential order of eminence obtains between infinite and finite being. To establish the order of dependence also requires a longer inferential process.[94]

6. *Every being is either absolute or relative*

The disjunction of absolute and relative also presents difficulties due to the Trinitarian speculation.[95] While Scotus does not cite them explicitly as contradictory disjunctive transcendentals, he does clearly indicate the transcendental nature of both notions.[96]

That absolute and relative are coextensive with being in disjunction is clear from this statement: "It seems unintelligible that there be any singular entity which is neither an entity *ad se* nor *ad aliud*." [97] However, the relative and absolute are not so opposed that they cannot coexist in one and the same physical entity or *res*. They are, however, formally opposed, so that no relation is formally identical with its foundation, though it may be really identified with it.[98] Admitting the real distinction of

---

of the term, otherwise each of the Divine Persons would be imperfect because He lacked the precise formal personality of the others.)

[94] Note how Scotus arrives at the conclusion, "Unum infinitum sufficienter terminat dependentiam omnium finitorum, et specialiter primum a quo dependent." *Quodl.* q. 5, n. 9; XXV, 211b; *De Primo Principio*, passim.

[95] *Oxon.* 1, d. 26, q. un.: "Utrum personae divinae constituantur in *esse* personali per relationes originis vel per aliqua absoluta?" (X, 291-353).

[96] *Oxon.* 2, d. 1, q. 4, n. 15; XI, 111a: Dico quod nihil alicujus generis dicitur de Deo, sicut dictum est *dist.* 8. *primi*, et sicut absoluta ita relationes, quae formaliter dicuntur de Deo, non sunt alicujus generis, sed trancendentia et passiones entis in communi, quia quidquid convenit enti inquantum est indifferens ad infinitum et finitum, convenit ei prius quam dividatur in genera, et ita est transcendens.

[97] *Oxon.* 1, d. 28, q. 3, n. 2; X, 410ab: Proprietas ut proprietas est aliqua entitas, alioquin non constitueret aliquod ens; aut igitur entitas ad se, aut ad alterum, aut neutrum. Entitatem enim aliquam esse singularem, quae nec sit entitas ad se nec ad aliud, non videtur esse intelligibile; ergo oportet quod ista entitas formaliter, vel sit ad se, et tunc constituit personam absolutam; vel ad alteram, et tunc ut proprietas erit relatio.

[98] *Oxon.* 1, d. 26; q. un., n. 42; X, 332a: Respondeo 'quod refertur est aliquid excepto relativo,' hoc est, includit aliquod absolutum, quod est

some relations from their foundation, Scotus consequently admits the possibility of an accidental entity which is essentially relative.[99]

That every relation implies an absolute is founded on the nature of a relation. For a relation requires two terms (the *relata*) and a basis or foundation for the relation. Both foundation and terms must ultimately be absolutes.[100] Hence the essential order existing between the relation in so far as the latter is real and its corresponding absolute is primarily one of dependence.[101] A further problem arises when one attempts to determine the nature of this dependence. The relation seems to be a quasi-form and the foundation and terms the quasi-matter. The relation is not immediately the result of strict causality in the sense that it terminates a productive act. The relation, at least the *intrinsecus adveniens*, arises necessarily once the terms and foundation are placed.[102]

---

fundamentum relationis, *Oxon.* 2, d. 1, q. 4, n. 28, XI, 179a: Concedo quod nulla relatio sit idem formaliter fundamento, etsi realiter et per identitatem quandoque contineatur in eo.

[99] Confer *Oxon.* 2, d. 1, q. 4, n. 5ff; XI, 98ff.

[100] *Oxon.* 1, d. 26, q. un., n. 12; X, 302a: Omnis relatio primo terminatur ad absolutum. *Oxon.* 2, d. 1, n. 4, n. 17; XI, 116b-117a: Fundamentum ergo relationis est aliqua entitas formaliter non includens illam relationem formaliter, quia si formaliter includeret eam, non esset formaliter relatio ad aliud, sed ad se, quia fundamentum suum est formaliter ad se, cum quo ponitur formaliter idem. Nec posset esse fundamentum primum relationis, ad hoc enim esset quaerere de illa relatione prima in qua poneretur; non est ergo praecise aliqua relatio fundamentum alicujus relationis, quod etiam apparet in relationibus divinis, ubi est maxima identitas in fundamento, et tamen fundamentum non est formaliter relatio. *Rep. Par.* 1, d. 13, q. un., n. 10; XXII, 199b.

[101] *Oxon.* 4, d. 12; q. 1, n. 8; XVII, 543b-544a: Respectus est essentialiter habitudo inter duo extrema, et ideo sicut tollere terminum ad quem est respectus, est tollere vel destruere respectum, ita tollere illud cujus est respectus, est tollere respectum et destruere rationem respectus; non ergo, quia accidens respectivum est accidens, ideo requirit subjectum, vel fundamentum, sed quia respectus est respectus, ideo requirit cujus sit et ad quod sit, etiam in Divinis.

[102] *Oxon.* 3, d. 1, q. 1, n. 14; XIV, 40b.

Much of the difficulty in the analysis of relations comes from the fact that a theory of relations is independent and prior to metaphysics. When the notions are given a real and metaphysical application, it is difficult to determine whether or not the relations as such have any reality apart from the reality of the *relata*. A discussion of this hotly disputed question is beyond the scope of this investigation.

7. *Every being is either simple or composed*

A composite being is defined as a being consisting of parts actually united to form a whole. Simple being on the contrary is one which lacks parts.[103] The contradictory nature of the two differential notions is apparent. As such, they form one of the pairs of primary differences of being as such.[104] Since a composite being requires an actual union of the parts (*non ut divisis, sed ut unitis*), the notion of composition involves the additional notion of potentiality and act[105] mentioned above.[106] Actual composition is itself a perfection but not a pure perfection. It involves the notion of potentiality and mutual dependence and perfectibility of the parts of the composite. Hence it is obvious that what is simple is more perfect to the extent that it lacks this imperfection. It is also possible to establish an order of dependence between what is simple and composite,[107] in the sense

[103] *Quodl.* q. 9, n. 8; XXV, 383b-384a: Compositum non est illud quod est nisi ex partibus, et hoc non ut divisis, sed ut unitis, ut patet ex 7 *Metaphys.* respectu hujus syllabae *ab*, de *a* et *b*. *Oxon.* 1, d. 8, q. 3, n. 6; IX, 569b: Dico quod simplicitas est simpliciter perfectionis, secundum quod excludit componibilitatem vel compositionem ex actu et potentia vel ex perfectione et imperfectione. See also *De Primo Principio,* c. 4, concl. 1, pp. 64-65.

[104] *Oxon.* 4, d. 11, q. 3, n. 46; XVII, 429a: Ens et unum dividitur in simplex et compositum.

[105] *Oxon.* 1, d. 5, q. 2, n. 15; IX, 489b: Unitas compositi necessario est ex ratione actus et potentiae, sicut assignat Philosophus. *Ibid.* 2, d. 3, q. 6, n. 16; XII, 145a: Dico quod compositio potest intelligi proprie prout est ex re actuali et re potentiali vel minus proprie, prout est ex realitate et realitate actuali et potentiali in eadem re.

[106] *Supra* pp. 145-146.

[107] *Oxon.* 2, d. 3, q. 4, n. 17; XII, 110b: Dico, quod necesse est de non composito ut parte fieri vel produci compositum, vel ibitur in infinitum.

that what is composed of parts is dependent on its parts. The full application of this to the scholastic notion of extension would make an interesting study in itself.

8. *Being is either one or many* [108]

9. *Being is either the same or different, similar or dissimilar, equal or unequal to every other being* [109]

This list, if not exhaustive, includes the principal disjunctive transcendentals treated by Scotus. It should be noted that Scotus frequently refers to what we have called " contradictory disjunctives " as contraries. The reason for this is that the differential notions express as a general rule intrinsic modalities of being and that consequently, while one may speak of potency and act, substance and accident, and the like as simple realities, these notions are not *simpliciter simplices*. What is actually meant is a potential being, an accidental being, and so on. From a logical viewpoint, no composite concept is ever simply contradictorily opposed to another composite concept. Thus for instance, there are at least two alternatives to the concept *ens in actu,* namely *non-ens* and *ens in potentia.* But the differential elements themselves are contradictorily opposed to each other. Consequently, in reference to being they do constitute a contra-

---

[108] *Oxon.* 3, d. 1, q. 3, n. 3; XIV, 81ab: Unum et multa sunt opposita immediate dividentia ens... quae circa quodcumque includunt contradictionem, et non unum in ente est multa necessario. *Metaph.* 1, q. 1, n. 23; VII, 22a.

[109] *Metaph.* 4, q. 1, n. 8; XII, 149b: Idem et diversum condividunt totum ens. *Rep. Par.* 1, d. 19, q. 3, n. 5; XXII, 236b; *Oxon.* 1, d. 19, q. 1, n. 3; X, 170a: Dicit Philosophus 10 Metaph. quod 'Omne ens omni enti comparatum, est idem vel diversum,' ita quod omne ens omni enti comparatum est aequale vel inaequale. Sicut igitur fundamentum identitatis, aequalitatis et similitudinis hoc modo communiter sumptum, est ens in communi comparatum ad quodcumque ens in communi, ita et illae relationes sunt transcendentes licet non convertibiles, cum disjunctione tamen dividentes ens, sicut dividitur in necessarium et possibile. *Quodl.* q. 6, n. 5; XXV, 242ab: *Aequale* et *inaequale* non dicantur nisi secundum quantitatem. Quantitas aliquo modo convenit omni enti cujuscumque generis, et per consequens licet *magnum* et *parvum* secundum eum sint passiones propriae quantitatis, tamen translative accepta sunt transcendentia, et passiones totius entis.

diction. Scotus illustrates this in connection with the notions of substance (*ens non in subjecto*) and accident (*ens in subjecto*).

> For although 'white man' and 'man who is not white' do not contradict one another, however, in reference to man, 'white' and 'non-white' do contradict one another. In a similar manner with regard to being, 'to be in a subject' and 'not to be in a subject' contradict one another.[110]

The reason we have emphasized the contradictory character of certain differential notions in respect to being is that it facilitates the inference from one to another which is expressed in the law of the disjunction. To deny that all being is accidental implies immediately that some being must be non-accidental. It is extremely important, therefore, to analyze carefully the precise meaning of the much abused metaphysical terms, such as potency, act and the like, to determine in what sense they may be regarded as naming primary differences of being, as well as to determine the precise value of the axioms based upon such notions and the field of their legitimate application.

THE LAW OF THE DISJUNCTION

Scotus was clearly aware of the principle that obtains between the disjunctive transcendental attributes of being, of which St. Bonaventure had made such striking use in his proofs for the existence of God. He enunciated it clearly in what we have called the law of the disjunction.

> In the disjunctive attributes ... as a universal rule by positing the less noble extreme of some being we can conclude that the more noble extreme is realized in some other being.[111]

Although he makes this observation more or less in passing to prove a particular point, there is no doubt that it occupied a very important place in his metaphysics. Time and again he makes use of it to prove specific points. While it is unfortunate

---

[110] *Metaph.* 4, q. 2, n. 11; VII, 163b: Licet enim *homo albus* et *homo non albus* non contradicant, tamen circa hominem album et non album contradicunt; a simili contradicunt circa ens in subjecto, et non in subjecto.

[111] Confer note 15.

that he did not give an *ex professo* analysis of the principle and its demonstration, he has indicated the general lines along which such a proof is to be constructed. Furthermore, from what has been said, the essential order existing between the members of each pair of disjunctives is apparent. And therefore one can say that in the *De Primo Principio,* Scotus has actually demonstrated the principle through his analysis of the implications of an essential order.

When he speaks in the opening passages of this work of certain attributes of being which involve an essential order,[112] it is not at all unlikely that he has in mind specifically the disjunctives in contradistinction to the the simple convertible attributes. For in the passage where he enunciates the law of the disjunction, he contrasts the *passiones disjunctae* with the convertible transcendentals precisely on this score; the latter are so immediately bound up with being it is difficult to show that any order exists between them so that one can be used as the starting point for a demonstration of the other. In the case of the disjunctives, however, a definite order does exist, so that the more noble member can be demonstrated from the less noble.

From what has been said of the essential order above, it is easy to see that in all of the disjunctives expressly mentioned by Scotus, an essential order prevails. If an essential dependence cannot be demonstrated, at least a difference in relative perfection sets up an essential order of eminence (*excedens-excessum*). Scotus brings this out expressly since he says of the disjunctives elsewhere, that when being is divided by opposites one of the differences expresses perfection, the other imperfection.[113]

In fact, in the course of his analysis the majority of the disjunctive attributes of being are actually correlated. Besides the notions of prior-posterior, eminent-exceeded, dependent and ' that on which something depends ', cause-caused, Scotus mentions act-potency as an illustration of the order of eminence.[114] Simplicity

---

[112] *De Primo Principio* c. 1, p. 2.

[113] *Oxon.* 1, d. 2, q. 7, n. 6; VIII, 520a: Quando ens dividitur per opposita, alterum dividentium est perfectionis in ente, alterum imperfectionis.

[114] *De Primo,* c. 1, p. 3.

and composition are analyzed in reference to dependence;[115] unity is involved in every essential order for "*in ordine essentiali ascendendo itur ad unitatem et paucitatem; igitur statur in uno.*"[116]

It is not to the present purpose to give in any detail the argumentation developed in the *De Primo Principio.* Attention is merely called to the fact that in the analysis of the implications of an essential order given by Scotus in the opening chapter, adequate proof for the law of the disjunction can be found.

SCHEME OF THE DISJUNCTIVE TRANSCENDENTALS OF BEING

Attributes which in disjunction are coextensive with being.

Those which are primarily correlatives

Being either:

1. is prior or posterior
2. is a cause or is caused
3. exceeds or is exceeded

Those which are either materially or formally contradictories

Being is either:

1. actual or potential
2. independent or dependent
3. necessary or contingent
4. substantial or accidental
5. finite or infinite
6. absolute or relative
7. simple or composed
8. one or many
9. the same (equal) or different (not equal)

---

[115] *Ibid.* c. 4; p. 64ff.

[116] *Ibid.* c. 3, p. 58; compare this with the *quarta via* of St. Thomas, *Summa Theol.,* I, q. 2, a. 3, c.

CHAPTER VII

The Pure Perfections

The pure perfections occupy the fourth level of transcendentality, since they are neither simply convertible with being as a class nor do they properly constitute a coextensive attribute in disjunction. While it is true that any given perfection can be reduced to one member of a logical disjunction, for instance, rational and non-rational, or wisdom and non-wisdom, it is not always clear that such a disjunction can be regarded as a primary differentiation of real being. For the universe of discourse in these cases does not seem to be being *qua* being but *living being* and *rational being* respectively. But whether or not the pure perfections can all be reduced to one or several of the first three classes of transcendentals, it is still justifiable to consider them as a class since they have a distinct function in metaphysics as a theologic. How important this function must have been in Scotus' eyes becomes apparent if we remember that the very discussion of transcendentality from which the key-text was selected is primarily intended to justify the inclusion of such pure perfections as wisdom, goodness and the like among the transcendentals.

In this chapter, then, three points come up for discussion: 1) the nature of a pure perfection, 2) the properties of pure perfections, 3) the function of the pure perfections in metaphysics as a science of the transcendentals.

THE NATURE OF A PURE PERFECTION

In defining a simple or pure perfection, Scotus alludes to the *famosa descriptio* of St. Anselm.[1] According to the mind of the saint, says Scotus, a pure perfection (*perfectio simpliciter*), that is, a perfection in the unqualified sense of the term, is one which,

[1] *Monologium,* c. 15 (PL 158, 162-163).

in anything possessing it, it is better to have than not to have.[2] Or as he puts it elsewhere, " it is better in anything whatsoever than that which is not-it."[3]

This description, or rule of thumb, serves to distinguish the *perfectio simpliciter* from the *perfectio limitata*. The latter contains in its formal notion the idea of limitation or imperfection, e. g. matter, reasoning, composition, the nature of man, an angel, or for that matter any created nature. Even when taken in their highest possible degree (for instance, a most perfect angel) they contain only a limited amount of perfection. Consequently they are perfections only in a relative sense and not *simpliciter*. They are perfections only in relation to the being in which they are found.

But this rule of St. Anselm's requires clarification on two counts before it can be used. When a pure perfection is said to be " better in anything whatsoever than that which is not-it " (*in quolibet est melius ipsum quam non ipsum*), the meaning of " which is not-it " (*non ipsum*) and " in anything whatsoever " (*in quolibet*) must be correctly understood.

Non ipsum, says Scotus, must not be restricted to a simple contradictory negation of the perfection in question, for then any positive perfection would be a pure perfection and a limited perfection would simply not exist. For the contradictory of a positive entity being a pure negation or non-entity (*nihilum*), any perfection, such as materiality, human nature, and the like, is better than no perfection or entity at all.[4] *Non ipsum* is rather

---

[2] *Quodl.* q. 5, n. 13; XXV, 216a: Secundum quod colligitur ex intentione Anselmi *Monol.* 15 perfectio simpliciter est quae, in quolibet habente ipsam melius est ipsam habere quam non ipsam habere. Confer also *Oxon.* 1, d. 2, q. 7, n. 39; VIII, 588a; *Oxon.* 1, d. 8, q. 1, n. 7; IX, 570a; *Ibid.*, d. 26, q. un., n. 36; X, 325a.

[3] *De Primo Principio,* c. 4 (Mueller ed.), p. 68: Perfectio simpliciter dicitur, quae in quolibet est melius ipsum, quam non ipsum.

[4] *Quodl.*, q. 5, n. 13; XXV, 216a: Haec autem regula indiget duplici expositione. Non enim intelligitur sic, melius est ipsum, quam sua negatio contradictorie opposita, quia sic quodlibet positivum esset perfectio simpliciter, quia quodlibet positivum est simpliciter melius sua negatione contradictorie opposita, sed intelligitur ibi non ipsum pro quocumque sibi

to be understood of anything that is incompatible with perfection in question, whether it be positive, negative or privative; whether it be opposed by way of contradiction or contrariety.

Take, for instance, any perfection, A. What is incompatible with it let us call B. Conceivably B may be simply a total lack of being (*nihil*), for nothing is not-A, or it may be a positive perfection or nature of which not-A may be predicated. For instance, if A is rationality and B the nature of a stone, B is incompatible with A, and to that extent, it is not-A. Since any kind of being is better than no being whatsoever, any perfection will be better than its negation in this sense. Hence Scotus immediately eliminates this interpretation of incompatibility, asserting that *non ipsum* is to be understood of *quodcumque positivum incompossibile.*[5] Whenever two positive perfections are incompatible with one another, if one is a pure perfection, it must be better than the other. For instance, human nature is incompatible with the divine, the angelic, the brute and the plant nature in the sense that as human nature, it cannot be a plant, a brute or an angel. Now while human nature is better than that of the plant and the brute, it is not better than that of the angel or of God, and therefore it is not better than anything whatsoever that is positively incompatible with it. On the other hand, intelligence, being, wisdom and truth are all better than anything incompatible with them. For example, what is intelligent is always better than what is not intelligent. So also, he who is wise is always better absolutely speaking than anything which cannot be wise.

In quolibet, says Scotus, is used rather than *cuilibet.*[6] For a pure perfection is not always better for everything (*cuilibet*). For instance, wisdom is not better for a dog, because it is incompatible with the nature of a dog. On the contrary, the perfec-

---

incompossibili etiam positive, ut sit sensus; perfectio simpliciter est in quolibet melior quocumque sibi incompossibili.

[5] *De Primo Princ.* c. 4, concl. 3, p. 69: Famosa est descriptio. Exponatur sic: melius quam non ipsum, id est, quam quodcumque positivum incompossibile, in quo includitur non ipsum.

[6] *Ibid.:* Est, inquam, sic melius in quolibet—non cuilibet, sed in quolibet —quantum esset ex se.

tion to be compared is something existing in something (*in quolibet*). Or as he puts it even more forcibly, *in quolibet habente ipsam melius est ipsam habere quam non ipsam habere.*[7] In other words, one perfection must be compared with a second incompatible perfection directly, rather than in reference to a particular nature which possesses the given perfection.[8] For example, rationality must be compared with anything that is incapable of being rational.

He brings this same point out elsewhere when he says that *in quolibet* refers not to a nature but to the supposit (*suppositum*).[9] *Suppositum*, as contrasted with nature, is understood in the traditional scholastic sense of concrete subsistence, that is, a " something-which-subsists." *Suppositum* or the supposit is to be taken in an absolute sense, without reference to the particular *kind* of nature which subsists.[10] If one were to ask whether it is better for a gold nugget to be gold or not to be gold, obviously a definitive answer could not be given. What is meant by a gold nugget? Is it primarily the supposit or the nature that is meant? If the nature of gold, or even the subsisting golden mass is meant, then for gold it is better to be gold than not to be gold. To be rational or living would destroy the nature of gold or this thing as a golden object. But if we

---

[7] *Quodl.* q. 5, n. 13; XXV, 216a.

[8] *De Primo Principio*, c. 4, concl. 3; p. 69: Breviter igitur dicatur; perfectio simpliciter est, quae est simpliciter et absolute melius quocumque incompossibili.

[9] *Quodl.* q. 5, n. 13; XXV, 216b: Secundo intelligitur *in quocumque*, non pro quacumque natura, sed pro quocumque supposito, non intelligendo, ut est talis naturae, vel talis, sed absolute accepto, ut est tale suppositum praescindendo rationem naturae, cujus est suppositum.

[10] *Oxon.* 1, d. 8, q. 1, n. 7; IX, 570b: Tunc debet intelligi hoc quod dicitur, in quolibet est melius, considerando quodlibet praecise inquantum suppositum, non considerando in qua natura illud suppositum subsistat; considerando enim aliquod inquantum subsistit in aliqua natura, potest aliqua perfectio simpliciter esse non melior sibi, quia incompossibilis ut est in tali natura, quia repugnat tali naturae, tamen ei inquantum praecise subsistens non repugnat; sed si hoc modo consideretur illam habere, erit simpliciter perfectius ens, quam si haberet quodcumque sibi incompossibile. *Oxon.* 1, d. 2, q. 7, n. 39; IX, 588a: *Quodl.* q. 5, n. 14; XXV, 216b.

abstract from the kind of nature, or even from the subsistence as determined to a definite nature, and consider instead, this nugget of gold merely in so far as it is something subsisting (*subsistens*), then it is much better for it if it were not an inanimate chunk of gold, but a living, rational thing.[11]

The attempt to justify and defend the description of St. Anselm has led Scotus to give a rather awkward and unwieldy definition of a pure perfection.[12] A simpler and more serviceable definition can be given in terms of some of the properties which Scotus deduces from this description based upon the *mens Anselmi.*

THE PROPERTIES OF PURE PERFECTIONS

Among the various properties of a pure perfection, there are four to which Scotus devotes special attention: 1) all pure perfections are mutually compatible; 2) pure perfections are compatible with the mode of infinity; 3) pure perfections, like perfection in general, are communicable in the sense that more than one person or supposit can share or participate in them;

---

[11] *Quodl.* q. 5, n. 13; XXV, 216b: Hoc patet quia auro sic considerato, ut habet naturam auri vel supposito considerato, ut subsistens in natura auri, non est melius non aurum quam aurum, quia incompossibile, ut incompossibile non est alicui melius; destruit enim entitatem ipsius; imo melius sic est auro esse aurum, quam quodcumque incompossibile naturae auri.

[12] Ockham, who follows Scotus in general on this point, has given a clearer formulation. *Ordinatio* I, d. 2, q. 3, P: Aliter accipitur perfectio simpliciter large et improprie pro omni conceptu ad quem consequentia formali non sequitur illud esse imperfectum; de quo illud verificatur sicut non sequitur formaliter A est sapiens, ergo A est imperfectum. Nec sequitur A est bonum, ergo A est imperfectum et sic de multis talibus. Et tales perfectiones simpliciter sunt multae et sunt attributa divina et possunt competere creaturae et sunt conceptus quidam communes deo et creaturae. Et sic intelligit Anselmus quod in quolibet est melius ipsum quam non ipsum, quia scilicet ex tali non sequitur formaliter illud esse imperfectum de quo verificatur et ex quolibet sibi incompossibili sequitur formaliter ipsum esse imperfectum. Sicut ex hoc quod A est sapiens non sequitur formaliter quod A est imperfectum et ex quolibet cui repugnat formaliter esse sapiens sequitur formaliter ipsum esse imperfectum de quo verificatur et ita aliquid esse melius.

4) they are irreducibly simple or *simpliciter simplices*. To explain further:

1. *All pure perfections are mutually compatible:* Were such not the case, then one and the same perfection would be both better and worse than itself or than any other pure perfection. Scotus argues as follows: If there could be two pure perfections which are mutually incompatible, let them be designated by A and B respectively. By definition,[13] A as a pure perfection must be better than B, which is incompatible with it. In like manner, B as a pure perfection will be simply and absolutely speaking better than A. But such mutually contradictory propositions cannot be true, for then any given pure perfection could be both inferior and superior to itself. But if the consequent is false the antecedent must be false likewise. Therefore every pure perfection must be compatible with every other pure perfection.[14]

2. *Every pure perfection is compatible with infinity.* This presupposes two propositions which must be established, but which we will pass over here: first, that infinity is a positive perfection or mode of being,[15] and secondly, that nothing can exceed that which is infinite. If these be granted, it follows that if a pure perfection were incompatible with infinity, it must be better than that which is infinite; for a pure perfection by definition must exceed, or be better than, anything incompatible with it. But this is impossible in virtue of the second proposition.

---

[13] Confer note 8.

[14] *Quodl.* q. 5, n. 8; XXV, 211a: Nulla perfectio simpliciter est incompossibilis alteri perfectioni simpliciter.... Probatur per rationem perfectionis simpliciter, quia enim ipsa est melior in quolibet, quam non ipsa, hoc est quam quodlibet incompossible sibi, sicut exponitur dictum Anselmi in ratione secunda principali hujus articuli; si sint ergo duae perfectiones simpliciter incompossibiles inter se, dicantur A et B, A erit in quolibet melius ipsum quam non ipsum, hoc est, quam quodlibet incompossibile sibi, et ita erit melius B, quod ponitur incompossibile sibi; et pari ratione B, si est perfectio simpliciter, erit melius quam A; talis circulatio est impossibilis, quia tunc idem esset imperfectius seipso.

[15] These propositions, which have come to be considered almost self-evident by Christian philosophers, were not so self-evident to the Greeks. Confer for instance Aristotle, *Physics,* III, 6.

Since the consequent is false, the antecedent must be false and its contradictory true. Every pure perfection, therefore, is compatible with infinity.[16]

The second property makes it possible to formulate a simple definition of a pure perfection. A pure perfection is any perfection that can be formally infinite.[17] Since whatever is infinite is by definition unlimited and without imperfection, a pure perfection can be described as anything whose formal notion contains as such no imperfection or limitation.

3. *Every pure perfection is communicable.* This third property is one of perfection as such rather than an exclusive property of pure perfections. The meaning is this, no person or supposit can so uniquely possess a given perfection that it can not be possessed by another. This proposition has special significance in the theological problem of the Trinity. Unless the notion of perfection be limited to quidditative or essential perfection, rather than extended to include the precise formal reality by which the Three Divine Persons are constituted in their respective personalities, one would have to admit that the Father would be imperfect because He lacks the perfection of being a Son. Similarly, the Holy Spirit would be imperfect because He lacks both Fatherhood and Sonship. St. Anselm, Scotus claims, has in mind only quiddities or essential perfections and not the personal properties of the Divine Persons when he described the pure perfections as he did.[18]

Scotus argues for the communicability of the pure perfections on theological grounds, using the divine nature as a specific instance, for it contains all pure perfections.[19] If the di-

---

[16] *Quodl.* q. 5, n. 9; XXV, 212a.

[17] *Oxon.* 1, d. 26, q. un., n. 37; X, 326a: Solum illud est perfectio simpliciter quod potest esse in aliquo infinitum. *Oxon.* 1, d. 8, q. 2, n. 3; IX, 576a.

[18] *Oxon.* 1, d. 26, q. un., n. 37; X, 326a: Ad illud Anselmi *Monolog.* respondetur quod loquitur tantum de quidditatibus, non de proprietatibus hypostaticis.

[19] See for instance *De Primo Principio,* c. 4, conclusio 3, pp. 68-71; *Quodl.* q. 5, n. 9; XXV, 212a.

vine nature were not communicable in the sense that it could be shared by several Persons, the definition of pure perfection would not be verified and consequently none of the perfections which it contains would be pure perfections. For these perfections would not be simply more perfect to have than not have except for the divine nature alone. They would not be more perfect *in quolibet,* that is, in any subsisting thing whatsoever (considering it precisely as subsisting as explained above); for it would be incompatible with everything else, and that not merely considering other entities as having this or that nature but considering them simply as *supposita.*[20]

4. *Pure perfections are irreducibly simple.* If the pure perfections were not irreducibly simple perfections, but composed of two more simple perfections, the latter would themselves have to be pure perfections, otherwise the resulting perfection would not be a pure perfection. Furthermore, they would have to be so united as to form an *unum per se,* for an accidental unity would certainly not be consistent with pure perfection. But in order to constitute a true *unum per se,* the constitutive perfections would have to be related to each other as act to potency. But this would mean that they are either physical or metaphysical parts and consequently imperfect and mutually perfectible. Hence they could not be pure perfections in themselves nor constitute a pure perfection.[21]

[20] *Oxon.* 1, d. 2, q. 7, n. 39; VIII, 588a: Omnis perfectio simpliciter est communicabilis pluribus...Probo, perfectio simpliciter est quae melius est ipsum esse quam non ipsum. Anselmus *Monol. cap. 15.* quod sic intelligitur, quod perfectio simpliciter est melius, quocumque sibi incompossibili in quolibet supposito absolute considerato, hoc est, non determinando in qua natura sit illud subsistens. Sed si natura divina determinaret se ad subsistentiam incommunicabilem, ipsa in nullo esset melior quocumque sibi incompossibili, nisi illa subsistentia ad quam se determinaret, quia cuilibet alii subsistentiae esset incompossibilis; igitur non esset perfectio simpliciter.

[21] *Quodl.* q. 1, n. 4; XXV, 10a: Omnis perfectio simpliciter est simpliciter simplex. Probatur, si est aliquo modo resolubilis in distincta, sint A et B, neutrum potest esse perfectio simpliciter simplex, quia tunc unum non faceret per se unum cum reliquo, quia non est per se unum ex quibuscumque distinctis, nisi unum sit sicut actus et aliud sicut potentia.

Scotus' analysis of the pure perfections emphasizes particularly the characteristic of compatibility. Where two perfections are mutually incompatible, the question naturally arises, Are they perfections in the unqualified sense of the term if, in perfecting their respective subject, they exclude the other perfection? Either we are forced to say that neither the one nor the other is more perfect and that such a thing as a pure perfection does not exist, or we must say that one of the two involves some imperfection. Therefore, to exclude it does not imply imperfection but rather perfection. By the analysis of the various perfections in the universe, it is possible to discover certain perfections which do not exclude of their very nature other perfections except where the latter evidently involve limitation and defects. This approach of Scotus is especially interesting since it seems to hold possibilities of providing a solution to an objection which contemporary thought would naturally raise. How can we determine whether or not one thing is better absolutely speaking than another? This is a point, however, that would demand further development and discussion than can be devoted to it here.

A further problem arises. Since the pure perfections do not involve in their formal concept any notion of imperfection, can any order exist among them? The definition of a pure perfection does not necessarily argue that all pure perfections are equally perfect. Being mutually compatible, it is not necessary, nor is it possible, for each to be better than the other. Scotus suggests that an order does exist among them if one considers merely the formal notion of each. The instance of the much-disputed scholastic question of the primacy of intellect or will comes to mind. But while it may be possible to admit some order on this basis, Scotus reminds us that if any given perfection is taken in the highest possible degree, we can no longer speak of it being more or less perfect than a second taken in its highest degree. All are equally perfect since they are formally infinite.[22]

---

[22] *Oxon.* 1, d. 8, q. 1, n. 8; IX, 570b: Ad secundum dubium dico quod requirit declarationem quis sit ordo perfectionum simpliciter, et modo breviter supponatur, quod aliquis ordo perfectionis sit inter eas, ita quod una ex ratione sui est perfectior alia praecise sumpta; licet quando quaelibet est in summo, sint tunc omnes aeque perfectae, quia infinitae, quaelibet tunc est infinita, de hoc alias dicetur.

THE FUNCTION OF THE PURE PERFECTIONS IN METAPHYSICS

As has been mentioned, it was the Scotus' desire to justify the transcendental character of the pure perfections that occasioned his discussion of the nature of transcendentality in our key-text. How can any notions drawn from creatures be predicated of God? Through his analysis of the problem of analogy, it became apparent that for an Aristotelian all analogous predication in regard to God presupposes univocal notions. On this basis, it becomes possible to discover transcendental elements in the data drawn from experience. By piecing such transcendental concepts together the human mind obtains a valid though imperfect knowledge of the transcendent God. It is because of their theological implications that Scotus attaches such importance to the pure perfections. To show that they are found *unitive* in God, the Supreme Being, is his ultimate goal.

In so far as the pure perfections form a class of their own, they have a distinct function to perform in this task. The notion of being as being is the starting point in such speculation. But because of its commonness, it can give us no distinctive or characteristic notions about God. It must be enriched from without. Through the disjunctive transcendentals, it is possible to arrive at the knowledge of a Being, first in the order of efficiency, finality and eminence. But this procedure only sketches the barest outlines. It is important to fill in the figures, to give positive content to the modalities of necessity, actuality, infinity and the like. This is done by applying the pure perfections. With the disjunctive transcendentals, the metaphysician singles out the imperfect elements in creatures and argues to a lack of such imperfection in the opposite extreme of the disjunction. Here, however, the metaphysician picks out the perfections of creatures, searching for those univocal elements which can be predicated of the *Primum Ens*, once the mind has prescinded from the mode of limitation associated with their actual state of existence in creatures.

It is not surprising, then, that in the *De Primo Principio* after having established the existence of God, Scotus follows up by setting up a principle which will enable him to demonstrate the intelligence and free will of God. This principle reads as follows:

> Every pure perfection is predicated of the highest nature as necessarily existing there in the highest degree.[23]

Once this principle is established, it is only necessary to analyze any given perfection to discover whether or not its formal notion implies limitation and one can immediately conclude whether it can exist in God or not.

It is not our intention to enumerate the series of theorems which have led up to and prepared the ground for this principle. The general tenor of his argument in this particular instance, however, can be indicated.

An essential order, he points out, exists between a pure perfection and anything incompatible with it, so that the pure perfection is by definition the more noble and consequently the *excedens* in the *ordo eminentiae.* Whatever is incompatible with it is always the less perfect or the *excessum.* But if the pure perfection is incompatible with the Supreme Nature, it would have to be more perfect than the most perfect nature—which is absurd. Hence it must be compatible with it. But if it is compatible with the Supreme Nature, it can pertain to the nature either as an accident pertains to its respective subject or necessarily. Now an analysis of the nature of a pure perfection reveals that it not only is compatible with the Supreme Nature as such, but it is compatible with a *per se* or necessary inherence in that nature so that it is either simply identical with the nature (in which case it is predicated *per se primo modo*) or at least it is a proper attribute (*propria passio,* and hence predicated *per se secundo modo.*). This compatibility of every pure perfection with necessary inherence in the Supreme Nature is proved as follows. Whatever possesses a given perfection, possesses it more perfectly if it does so necessarily than if it does so contingently, providing that necessity of inherence is not repugnant to the perfection in question. But necessary inherence is not repugnant to a pure perfection, for if it were, a pure perfection

---

[23] *De Primo Principio,* c. 4, conclusio 3, p. 68: Omnis perfectio simpliciter, et in summo, inest necessario naturae summae.

would be exceeded by something incompatible with it, namely by a perfection which does necessarily inhere or can necessarily inhere—contingent inherence implying dependence (causality) and therefore imperfection on the part of the attribute which inheres. Now no nature can possess a given perfection more perfectly than the most perfect nature; hence we must conclude that it is possible for a pure perfection to inhere necessarily in the Supreme Nature. And if it is possible for the pure perfection in question to exist in the highest degree anywhere, surely it is possible as regards the Supreme Nature. And since all this is possible, argues Scotus, therefore *habetur propositum, quod necessario inest.*[24]

Scotus here makes use of the one instance where an inference *a posse ad esse* is valid, namely where the mode of necessity and the mode of incausability are predicated simultaneously. For he presupposes here a previously established principle, namely, *quod suprema natura est incausabilis.*[25] Since it is possible for the uncaused First Nature to possess a pure perfection necessarily, therefore it does possess it necessarily. This mode of reasoning is but a variation of the Aristotelian principle "in the case of eternal things, what can be must be."[26] To deny the validity of the inference is to affirm that the possibility of a pure per-

---

[24] *Ibid.* p. 70: Probo tertiam conclusionem sic intellectam; perfectio simpliciter ad omnem incompossibilem aliquem habet ordinem secundum nobilitatem, non excessi per descriptionem, sed eminentis; igitur vel est naturae supremae incompossibilis et ita excedit eam, vel compossibilis et ita potest illi inesse et etiam in summo, quia sic est sibi compossibilis, si est alicui compossibilis. Non autem inest ut accidens contingens; igitur vel ut passio propria saltem; habetur propositum, quod necessario inest.

Quod autem non ut accidens per accidens contingenter insit, probo: quia in omni perfectione, cui non repugnat necessitas, perfectius habet illam quod habet necessario quam quod contingenter; perfectioni simpliciter non repugnat necessitas, quia tunc aliqua incompossibilis sibi excederet eam ut illa quae est necessaria, vel potest esse. Nihil autem potest perfectius habere perfectionem simpliciter quam prima natura, ex secunda hujus; ergo, etc.

[25] *Ibid.*, c. 3, concl. 13; p. 53.

[26] *Physics* III, c. 4 (203b 30): ἐνδέχεσθαι γὰρ ἢ εἶναι οὐδὲν διαφέρει ἐν τοῖς ἀϊδίοις. *De Generatione et Corruptione* II, c. 11 (337b 35-338a 2).

fection inhering in the Supreme Nature is conditioned by something extrinsic to both and hence to affirm that the Supreme Nature, in regard to this perfection which it possesses, is or can be caused.

To sum up: Scotus has based his definition, or better, his description, of a pure perfection on that given by St. Anselm in the *Monologium,* but has clarified it by introducing the notion of compatibility. *Breviter igitur dicatur: perfectio simpliciter et absolute melius quocumque incompossibili.*[27] An analysis of this definition reveals the following properties: a pure perfection is compatible with every other pure perfection; a pure perfection can exist in an unlimited or infinite degree; a pure perfection is communicable; a pure perfection is irreducibly simple.

The pure perfections assume an important function as a distinct class only if metaphysics is regarded primarily as a theologic. While the disjunctive transcendentals, by reason of the essential order existing between the two members, are of paramount importance as the *medius fecundior* of establishing God's existence and limn the bare outlines in terms of existential or essential modalities, it is by an analysis of the pure perfections abstracted from the world of creatures that we are enabled to add body and substance to the notion of God. Not every pair of disjunctive transcendental attributes contains one member that can be applied properly or exclusively to God, for instance, substance and accident, one and many. But since one member of every true disjunction is a pure perfection, in many cases to establish the reality of such a perfection in God or to demonstrate that the first being in each of the various essential orders coincide in whole or in part, the simplest method is to make use of the " Principle of the Pure Perfections," namely, every pure perfection is predicated of the Supreme Nature as necessarily existing there in the highest degree. Scotus, incidentally, does use just this approach in proving that an infinite Being exists.

If we have singled out the pure perfections as a distinct class of transcendentals, one might wonder why we do not go on to speak of a fifth class comprising those notions which, though

[27] Confer note 8.

neither coextensive with being as such nor predicable of God, are nevertheless transcendental in character, for instance, such notions as contingency, finite, accident, and the like considered simply in themselves and not as part of a disjunction. Theoretically speaking, there is no reason why one might not go on to distinguish other layers of transcendentality, at least if we regard metaphysics as a science of being without any further qualification. But if one were to consider metaphysics primarily as a theologic (whose end and purpose is to give us knowledge of God) rather than as an ontology, further differentiation of transcendentality apart from any theologic implications would remain little more than barren speculation. But this leads to the final chapter.

CHAPTER VIII

CONCLUSION: METAPHYSICS AS A THEOLOGIC

It remains in conclusion to add a word on the relation of the theory of transcendentals to the metaphysics of Scotus as a whole. It is not too much to say that the theory of transcendentals is not merely an essential part—it is the whole of his metaphysics. As he put it in the prologue of his *Commentary on the Metaphysics of Aristotle:*

> It is necessary that some universal science exist which considers the transcendentals as such. This science we call "Metaphysics," from *meta,* which is "beyond" and *physis* [physical] "science." It is, as it were, a transcending science because it deals with the transcendentals.[1]

This conception of metaphysics as the "science of the transcendentals" raises many an interesting problem, such as whether the categories or predicamental being belong to metaphysics or physics, the relation of logic and metaphysics, the nature of the first subject of metaphysics, and so on. But in view of the strong theological coloring that suffuses Scotus' transcendentalism, there is one question that requires at least a word of comment—the relation of metaphysics and natural theology.

In the light of the researches of Jaeger,[2] Ross,[3] and others, we

---

[1] *Metaph.* prol. n. 5; VII, 5a: Necesse est esse aliquam scientiam universalem, quae per se consideret illa transcendentia, et hanc scientiam vocamus Metaphysicam, quae dicitur a *meta,* quod est trans, et *physis* scientia, quasi transcendens scientia, quia est de transcendentibus. *Ibid.*, n. 10; 7a: De isto autem objecto hujus scientiae, ostensum est prius, quod haec scientia est circa transcendentia.

[2] Werner Jaeger, *Aristotle,* trans. Richard Robinson (Oxford, Clarendon Press, 1934).

[3] W. D. Ross, Introduction to *Aristotle's Metaphysics* (Oxford, Clarendon Press, 1924) I, pp. xiii-xxxiii.

know today that Aristotle's *Metaphysica,* far from constituting an organic whole, is a more or less loose collection of metaphysical discussions interspersed with much matter of physics and logic. It is extremely doubtful whether Aristotle himself had a clear idea of the precise object of his " first philosophy." In the opening chapter of the fourth book we have a striking evidence of his uncertainty of mind. Is metaphysics a special science or is it universal? Does it deal with one particular class of real beings—the " immovable substances," which are God and the " Intelligences "—or is it concerned with the common aspects of *all* real beings? Is it primarily a natural theology or an ontology?

His uncertainty of mind only became accentuated in the two Arabian strains of Aristotelianism. Avicenna claimed that being *qua* being is the subject of metaphysics; Averroes, that it is God and the Intelligences. And so the problem of the subject or object of metaphysics was bequeathed to the medieval schoolmen, a problem which Scotus takes up in the opening question of his *Metaphysics.*[4]

It is not our intention to determine which view Scotus finally adopted and with what modifications. This question cannot be definitely decided without more information regarding the chronology of Scotus' writings. It should be noted, however, that the question requires a deeper and more penetrating study than has hitherto been accorded it. While the importance of Gilson's " Avicenne et le point de départ de Duns Scot "[5], a pioneer work in this field, should not be underestimated, its conclusions can not be regarded as final. To say that Scotus, after some hesitation, finally sides with Avicenna seems somewhat of an oversimplification of his real position. It overlooks the great labor and the extreme care that Scotus devotes to a defense of Averroes' essential position, that God is the subject of metaphysics,

---

[4] *Metaph.* 1, q. 1; VII, 11: Utrum subjectum Metaphysicae sit ens inquantum ens sicut posuit Avicenna? vel Deus, et Intelligentiae, sicut posuit Commentator Averroes? Note that the formulation of the question in the Vivès edition is defective.

[5] E. Gilson, " Avicenne et le point de départ de Duns Scot " in *Archives d'Histoire doctrinale et littéraire du Moyen Âge,* II (1927), 89-149.

but *aliter est ponendum quam ponit Averroes.*[6] If Scotus broke with Averroes, it was not because the Commentator made a theology of his metaphysics. It was because Averroes had relegated to physics the task of proving the existence of God and had thus made metaphysics dependent upon physics for a knowledge of its first subject.[7] And furthermore, just what did Scotus mean by the subject and object of a science in general? How did he apply this notion to metaphysics? A great deal more research will be required before these questions can be adequately answered.[8]

---

[6] *Metaph.* 1, q. 1, n. 34; VII, 28a: Tenendo quod Deus sit hic subjectum, aliter est ponendum quam ponit Averroes.... Primo enim ostendetur quomodo peccavit Averroes et Avicenna in opinionibus suis. Secundo dicetur modus quo Deus potest poni subjectum in Metaphysica.

[7] Gilson himself calls attention to this point, *op cit.*, p. 98. Confer Scotus, *Metaph.* 1, q. 1, n. 7; VII, 14b: Scientia naturalis simpliciter erit prior ista [sc. Metaphysica]. *Ibid.*, 6, q. 4, n. 3; 349a.

[8] Since the *Quaestiones Subtilissimae super lib. Arist. Metaphys.* contain an *ex professo* treatment, whereas the passages in the *Opus Oxoniense,* etc. are *obiter dicta,* the latter should be interpreted in the light of the former and not vice versa, unless it can be definitely shown that the Metaphysics represents an earlier work. Gilson's interpretation is open to criticism on this score. He assumes that the *Metaphysics* " n'est pas le travail d'un théologien; mais c'est le travail d'un maître ès-arts qui a déjà pris contact avec la littérature théologique et peut se permettre publiquement un jugement sur Henri de Gand ou frère Thomas d'Aquin." (*op. cit.*, p. 92, note 1.) It is not clear on what basis he makes this assertion. If the unfinished condition of the *Metaphysics* is any indication, it would seem easier to argue for a later than for an earlier date of composition. Furthermore, there does not seem to be any general precedent to fall back upon which would indicate that in the 13th and early 14th century scholastics were wont to compose commentaries on the works of Aristotle prior to their commentaries on the *Sentences* of Peter Lombard. If anything the converse seems true. St. Thomas, for instance, did not compose his commentary on Aristotle's *Metaphysics* until well over a decade after he had completed his work on the *Sentences.* (Confer Grabmann, " Die echten Schriften der hl. Thomas v. Aq." in *Beiträge zur Geschichte der Philosophie und Theologie des Mittelalters,* XXII, 1-2 (Münster, 1920).) Boehner has definitely established a similar sequence for Ockham. (*The Tractatus De Successivis Attributed to William Ockham* (St. Bonaventure, Franciscan Institute, 1944), p. 16; also the article on the *Ordinatio* of Ockham in *The New*

If Scotus is not too certain whether the subject of a science is to be determined primarily from the standpoint of origin or end,[9] or whether any metaphysics we possess can be called a *scientia propter quid*,[10] one thing is quite clear—his opposition to any divorce of natural theology from metaphysics.[11] This position he has not abandoned in any of his works. Whether God be considered the primary object or not, so much is certain, God is *an* object of metaphysics, for He is " considered in this science in the most perfect way it is possible for Him to be considered by any naturally acquired science." [12]

For that reason Scotus fought against the introduction of a fourth speculative science,[13] namely, an ontology, which treats the common elements abstracted from God and creatures, while God is relegated to a special theologic which treats of God as God and is on a par with the other special speculative sciences of physics and mathematics. Such a division might be justified

---

Scholasticism, XVI (1942), 223). Most probably the *Metaphysics* and the *Opus Oxon.* were being worked on simultaneously, since mutual cross references characterize the two works.

[9] Confer, for instance, the prologue to the *Metaphysics* (n. 10; VII, 6b-7a) where he points out that the subject (or more properly the object) of a science partakes of the character of all four causes. While it is relatively easy to define the subject of a *scientia propter quid* where these various aspects are found in one and the same object, in a *scientia quia* this is not true. The starting point has the character of efficient causality, whereas the object whose existence is to be proved has the character of final causality.

[10] Confer, for instance, *Metaph.* 1, q. 1, n. 36; VII, 29a. The fact that the attributes cannot be deduced from the concept of being also makes it impossible for metaphysics even as a science of *ens qua ens* to be a strict *scientia propter quid.*

[11] *Metaph.* 1, q. 1, passim, esp. nn. 34-49; VII, 28a-37a; *ibid.* prol. n. 10; 6b-7a; *Rep. Par.* prol. q. 3, n. 3; XXII, 47b, etc.

[12] *Oxon.* prol. q. 3, n. 20; VIII, 171a: Deus vero, etsi non est subjectum primum in Metaphysica, est tamen consideratum in illa scientia nobilissimo modo quo potest in aliqua scientia considerari naturaliter acquisita.

[13] *Metaph.* 1, q. 1, n. 48; VII, 36a; *ibid.* n. 49; 36b: Ideo vitando quatuor esse scientias speculativas, et hanc ponendo de Deo omnia naturaliter cognoscibilia de ipso sunt transcendentia.

if God could be apprehended by direct intuition and studied through proper notions, or even if we possessed a natural science other than metaphysics which would give us adequate knowledge of Him.[14] But in the present state of existence, all our natural knowledge of God is in terms of transcendentals.[15] Consequently, natural theology lies within the realm of metaphysics, which by definition is the science of the transcendentals.

Furthermore, if the task of metaphysics as a science of being *qua* being is to demonstrate—if demonstration be possible—or to establish not only the proper attributes which are simply convertible, but also the disjunctive attributes, as Aristotle pointed out,[16] or where only one member of the disjunction is given in experience, to establish the other, then how is it possible to divorce ontology from its role as a theologic without either reducing it to a sort of glorified taxonomy, if it be permitted to discuss the interrelation and implication of such notions, to keep it from passing over into the realm of pure logic? For so long as metaphysics stays in the real order, though it begin with the notion of being *qua* being, it will invariably end with the notion of God. It cannot analyze the notion of contingency save in terms of necessity, or the relative without introducing the absolute, or order without a primacy. As the metaphysician goes on analyzing the implication of each basic metaphysical notion, he unconsciously adds a new line to what proves to be a portrait of God. As Scotus beautifully expressed it, only because God is pure Act, is being divided into act and potency.[17] Thus the science of the *ens in communi* finds its consummation in the discovery of the *primum ens*. For Scotus our whole ontology is

---

[14] While Scotus admits that physics can prove the existence of God, the knowledge so derived is very imperfect, namely, of God as the First Mover. Even here, the physicist must have recourse to metaphysics to prove that his First Mover is really first. Confer *Metaph.* 6, q. 4, n. 2; VII, 348.

[15] *Metaph.* 1, q. 1, n. 49; VII, 36b.

[16] Aristotle, *Metaphysica* IV, c. 2 (1004a 9ff); Scotus, *Metaph.* 1, q. 1, n. 48; VII, 36ab.

[17] *Metaph.* 1, q. 1, n. 43; VII, 33b: Quia enim Ipse est actus purus, et ideo ens dividitur per actum et potentiam.

saturated with theologic. What Scotus has actually attempted in the opening question of his *Metaphysics* is to synthesize Avicenna and Averroes, or better, to weld the heterogeneous elements of Aristotle's *Metaphysics* into an organic whole. And if this be true, must not *De Primo Principio* be considered in a new light? Has he not sketched here the rough outlines of what he himself considered to be an ideal scientific metaphysics?

In the philosophy of Duns Scotus, then, the theory of transcendentality assumes a much more important role than that usually assigned to it in contemporary scholastic philosophy. Adopting the definition of Albert the Great, Scotus reveals the richness and extent of the realm of the transcendentals. Any real notion that rises above the predicamental sphere and that cannot be confined to a single category or supreme genus deserves the name of transcendental. Four main classes are set apart: the transcendental notion of being; the coextensive or simple convertible attributes, such as one, good and true; the attributes of being coextensive in disjunction, such as infinite-finite, act-potency, necessary-contingent; and the realm of the pure perfections, such as substance, life, freedom, intelligence, wisdom and the like.

These metaphysical notions are real concepts and, as such, are to be distinguished from the sphere of logical intentions, for logic has its own transcendental notions. The metaphysical transcendentals are predicable immediately of real beings. While the single transcendental perfections do not necessarily represent the totality of the physical thing, they do represent certain real aspects of things. The objective reality of such real, though partial, concepts is guaranteed by a formal or at least a modal distinction *a parte rei*. The latter distinction is extremely important for the theory of transcendentality. It is by abstracting from the intrinsic mode of finiteness, which characterizes all creature perfections, that the mind forms its most important transcendental notions, namely, the notions of pure perfections which are common to God and creatures. These common, imperfect and univocal notions must be distinguished from the proper composite concepts of God or of substance through which the mind conceives the formal perfection together with its instrinsic mode.

Being is the first and most important of the transcendental notions. As a term, however, being is equivocal. It may designate at times the total entity of a thing. That is to say, it may signify that by which a thing is both like and unlike another. When the common term of "being" is used to designate everything according to the proper entity of each, it is predicated analogously or equivocally. But in this case there is no true unity of concept, but there are as many concepts as different types of being. Furthermore, such concepts are not irreducibly simple, for it is always possible to prescind from the differential elements and form a common univocal, though imperfect, concept. This latter is what is truly transcendental being. It is an irreducibly simple concept. It represents the ultimate determinable element in everything that is not primarily diverse but only different. Its content is expressed by the phrase, " that to which existence is not repugnant " (*cui non repugnat esse*). It is primarily a quidditative or essential notion, but a quiddity defined in terms of existence. This notion applies primarily to being understood in a nominative sense, that is, *a being*. In a derivative sense (denominative) it can be applied to those notions which express the various qualifications or differential notions. Being is said to have a twofold primacy in regard to all intelligible notions. It is predicable univocally as the common determinable note or *quid* of all physical things (*res*) or of any reality that is grasped by a concept that is not irreducibly simple. It enjoys a primacy of virtuality in regard to all other irreducibly simple qualifications, whether they be attributes or ultimate differences. Being is said to contain the latter virtually, not in the sense that these other notions can be inferred or derived from the concept of being, but in the sense that they are contained *in beings*, that is, in those things of which the term " a being " is predicable essentially.

The coextensive attributes of unity, truth and goodness are not convertible in the sense of being formally identical or synonymous with being. The meaning is simply this: whatever is a being is also one, true and good. Or to put it in more technical terms, the convertible transcendentals are predicable *in quale* of everything of which being is predicable *in quid*. As irreducibly simple qualifications (ultimate *qualia*) they lie outside the concept of being considered precisely as a subject of further determination, that is, being considered as the ultimate *quid*. To this extent they are formally distinct both from each other and from being. Because of the predominant theological coloring of Scotus' theory of transcendentality, these coextensive attributes occupy only a minor place in the theory as a whole.

Far more important are the disjunctive transcendentals. They fall into two main classes, the correlatives, like prior-posterior, cause-caused, and those which are contradictorily opposed, as infinite and finite. The analysis of the mutual relations and implications of the disjunctive transcendentals is in reality the principal task of the metaphysician and the conclusions arrived at in this way form the main body of metaphysics as a science. These disjunctives are not deduced from being any more than are the notions of true, goodness, and the like, though such propositions as "Being is either finite or infinite" are immediately evident (*per se notae*). The notions of the imperfect members of the disjunction are the fruit of experience, though, it should be noted, the barest minimum of experimental data—whether internal or external—suffices. In those disjunctions contradictorily opposed, the existence of the imperfect member implies the existence of the more perfect member but not vice versa. With the correlatives the existence of either member implies the other.

If the fourfold division of the transcendentals were to be considered as an adequate division, the class of pure perfections would not really form a distinct coordinate member, for being, the coextensive transcendentals and the more perfect member of each disjunction are also regarded as pure perfections. But the realm of pure perfections is much broader. Pure perfections that are proper to God, such as pure act, are only grasped by us in composite concepts. Hence, so far as we are concerned, such concepts are secondary transcendentals. Scotus is more interested in those simple concepts of pure perfections univocally common to God and creatures.

The theory of transcendentals is not simply an important section of Scotus' metaphysics. It is his metaphysics. Like his metaphysics, it is saturated with theological implications. If it begins with being *qua* being it does not rest until it reaches its goal in God the ultimate efficient and final cause, with the knowledge of whom *terminatur scientia metaphysicalis.*[18] Beginning with the simple univocal notion of being, the metaphysician goes on to analyze the various conditions of being as it actually exists, namely, its contingency, its limitation, its multiplicity, composition and the like. Through the law of the disjunctive transcendentals he rises to a knowledge of the more perfect member of each disjunction. Further analysis reveals that these higher attributes must coincide in a being which is first in the order of eminence, finality and efficient causality. Through the law of the pure perfections it is possible to bring out the highlights of this being we call God, until He is revealed to be the ultimate solution to the fundamental question of the metaphysician: Why does being exist? In this sense, metaphysics is a truly existential science, and the theory of the transcendentals a genuine θεωρία—a contemplation of God.

---

[18] *Rep. Par.* prol. q. 3. n. 3; XXII, 47b.

BIBLIOGRAPHY

PRIMARY SOURCES

Joannis Duns Scoti Opera Omnia, 26 vols. Vivès edition, Paris, 1891-1895.

Commentaria Oxoniensia, 2 vols. Fernandez Garcia edition, Quaracchi, Collegium S. Bonaventurae, 1912-1914.

Quaestiones Quatuor scripti Oxoniensis super Sententias, 3 vols., ed. Salvator Bartoluccius. Venetiis, apud Haeredes Melchioris Sessae, 1580.

Commentaria Oxoniensia in quatuor libros Sententiarum et Quodlibeta. ed. Thomas Penketh. Nuremberg, Anthony Koberger, 1481.

Quaestiones subtilissimae super libros Metaphysicae Aristotelis, Venetiis, Joannes Hertzog, 1499.

Quaestiones aureae et subtiles super libros Elenchorum Aristotelis, Venetiis, Joannes Hertzog, 1495.

Collationes, ed. Harris, *Duns Scotus,* Oxford, Clarendon Press, 1927, vol. II, pp. 361-378.

Collationes, ed. Balić in "De Collationibus J. Duns Scoti," *Bogoslovni Vestnik,* IX (1939), 185-219.

"Une question inédite de Jean Duns Scot sur la volonté" by Balić in *Recherches de Théologie Ancienne et Médiévale,* III (1931), 191-208.

Tractatus de Primo Principio, ed. Marianus Mueller. Freiburg, Herder, 1941.

SCOTISTIC COMMENTATORS

Andreas, Antonius, *In XII libros Metaphysicae Aristotelis,* Venetiis, ex Haeredibus Octaviana Scoti, 1501.

Carolus-Josephus a S. Floriano, *Joannis Duns Scoti Philosophia,* 7 vols. Mediolani, ex typographia Marelliana, 1771-1783.

Cavellus, Hugo, *Commentaria,* found in the *Opera Omnia* Scoti, Vivès edition, passim.

Faber, Philippus, *Philosophia Naturalis Joannis Duns Scoti,* Parma, 1601.

———, *Disputationes Theologicae,* Venetiis, ex officina Marci Ginammi, 1626.

Ferrari, Joannes, *Philosophia Peripatetica,* Venetiis, apud Modestum Fentium, 1746.

Frassen, Claudius, *Philosophia Academica,* Romae, ex typogr. Rocchi Bernabo, 1726.

———, *Scotus Academicus,* Romae, ex typographia Sallustiana, 1900.

Gadius, Hieronymus, *Lectura in Quodlibetum J. Scoti,* Bononiae, J. Bapt. Phaellus, 1533.

Gemmelli, Julius, *Duodecim mirabiles gyri ad mentem Platonis, Aristotelis, et Theologum, totum cursum scholasticum Theologiae, Philosophiae et Logicae, circumgyrantes . . .*, Venetiis, apud Robertum Meiettum, 1592.

Henno, Franciscus, *Theologia Dogmatica, Moralis et Scholastica,* Venetiis, apud Antonium Bortoli, 1719.

Joannes Canonicus, *Quaestiones super VIII Libros Physicorum Aristotelis,* Venetiis, Octavianus Scotus, 1481.

Lychetus, Franciscus, *Commentaria,* found in the *Opera Omnia Scoti,* Vivès ed., passim.

Mastrius, Bartholomeus, *Cursus Philosophicus,* Venetiis, apud Nicolaum Pezzana, 1708.

———, *Disputationes Theologicae in Primum Librum Sententiarum,* Venetiis, apud Joannem Jacobum Hertz, 1698.

Mauritius de Portu, *Commentaria,* found in *Opera Omnia Scoti,* Vivès ed., passim.

Mayron, Franciscus, *Super Primum Librum Sententiarum,* Basilee, per Nicolaum Kestler, 1489.

———, *Super Secundum, Tertium, et Quartum Librum Sententiarum,* Venetiis, apud Haeredes Octaviani Scoti, 1505-1507.

———, *Quaestiones Quodlibetales,* Venetiis, apud Haeredes Octaviani Scoti, 1507.

Montefortino, Hieronymus, *Summa Theologica,* Romae, ex typographia Sallustiana, 1900.

Petrus de Aquila, *Commentaria in IV Libros Sent. Magistri Petri Lombardi,* ed. Cyprianus Paolini, Levanti, Conv. SS. Annuntiationis, 1907.

Poncius, Joannes, *Integer Philosophiae Cursus ad mentem Scoti,* Parisiis, Antonius Bertier, 1648.

Sanning, Bernardus, *Scholae Philosophicae Scotistarum,* Neo-Pragae, typis Hampelii, 1685.

———, *Schola Theologica Scotistarum,* Vetero-Pragae, typis Danielis Michalek, 1679.

Tataretus, Petrus, *Universae Aristotelicae Logicae Explanatio,* Venetiis, apud Haeredes Melchioris Sessae, 1571.

Vallo, Joannes, *Lectura Absolutissima in Formalitates Scoti,* Venetiis, apud Franciscum de Franciscis Senesem, 1588.

OTHER WORKS

Albanese, Cornelio, "Intorno alla nozione della verità ontologica," *Studi Francescani,* XII (1913-1914), 274-287.

Albert the Great, St., *Opera Omnia,* ed. A. Borgnet, Paris, Vivès, 1890-1899.

Alexander of Hales, *Summa Theologica,* Quaracchi, Coll. S. Bonaventurae, 1924-30.

Aristotle, *Opera Omnia*, ed. I. Bekker. Berlin, 1831-1870.

———, *Metaphysics*, ed. W. D. Ross. Oxford, Clarendon Press, 1924.

Augustine, St., *De Trinitate Libri quindecim*. Migne edition, *Patrologiae Cursus Completus*, Series Latina (PL) v.42, 819-1098.

———, *Epistola 147 De Videndo Deo, ad Paulinam*, PL 33, 596-622.

Averroes, *Commentaria*, found in the Iuntas edition of *Aristotelis opera omnia*, Venetiis, 1553.

Avicenna, *Opera*, trans. Dominicus Gundissalinus, Venetiis, 1508.

Balić, Carl, " A propos de quelques ouvrages faussement attribués a Jean Duns," *Recherches de Théologie Ancienne et Médiévale*, II (1930), 160-188.

———, " De Collationibus Joannis Duns Scoti," *Bogoslovni Vestnik*, IX (1939), 185-219.

———, *Les Commentaires de Jean Duns Scot sur les quatre livres des sentences*, (Bibliothéque de la Revue d'Histoire Ecclésiastique, Fasc. I) Louvain, Bureaux de la Revue, 1927.

———, *Relatio a Commissione Scotistica exhibita Capitulo Generali Fratrum Minorum Assisii A. D. 1939*, Romae, 1939.

———, *Ratio Criticae Editionis Operum Omnium J. Duns Scoti*, II (1939-1940) Romae, 1941.

———, " De Critica Textuali Scholasticorum Scriptis Accommodata," *Antonianum*, XX (1945) 267-308.

Barth, Timothy, " De fundamento univocationis apud J. D. Scotum " *Antonianum*, XIV (1939) 181-206; 277-298; 373-392.

———, " Die Stellung der univocatio im Verlauf der Gotteserkenntnis nach der Lehre des Duns Skotus," *Wissenschaft und Weisheit*, V (1938) 235-254.

Belmond, Séraphin, " Essai de Synthèse Philosophique du Scotisme," *La France Franciscaine*, XVI (1933) 73-131.

Bennet, Owen, *The Nature of Demonstrative Proof According to the Principles of Aristotle and St. Thomas Aquinas*, Washington, D. C., Catholic University of America Press, 1943.

Bettoni, Efrem, *Vent'Anni di Studi Scotisti* (1920-1940), (Quaderni della Revista di Filosofia Neoscolastica), Milan, Direzione e Redazione: Via Ludovico Necchi N. 2, 1943.

———, *L'Ascesa a Dio in Duns Scoto*, Milan, Societa Editrice " Vita e Pensiero ", 1943.

Binkowski, Johannes, *Die Wertlehre des Duns Skotus*, Berlin, F. Dümmler, 1936.

———, " Die Wertlehre des Duns Skotus in ihrer Bedeutung für die Gegenwart," *Wissenschaft und Weisheit*, III (1936) 269-282.

Boehner, Philotheus, *History of Franciscan Philosophy* (manuscript).

———, " Medieval crisis of logic and the author of the Centiloquium attributed to Ockham," *Franciscan Studies*, XXV (1944), 151-170.

- ——, " The text tradition of the Ordinatio of Ockham," *The New Scholasticism,* XVI (1942) 203-241.

- ——, *Tractatus de Successivis Attributed to William Ockham* (Franciscan Institute Publications I), St. Bonaventure, Franciscan Institute, 1944.

——, " Scotus' Teachings according to Ockham. I. On the Univocity of Being; II. On the *Natura Communis," Franciscan Studies,* VI (1946), 100-107; 395-407.

——, and Étienne Gilson, *Die Geschichte der christlichen Philosophie,* Paderborn, Ferdinand Schöningh, 1937.

Boëthius, *In Categorias Aristotelis Libri Quatuor,* ed. Migne, PL 64, 159-294.

Bonaventure, St., *Opera Omnia,* Quaracchi, Coll. S. Bonaventurae, 1882-1902.

Borgmann, Pacificus, " Gegenstand, Erfahrungsgrundlage und Methode der Metaphysik," *Franziskanische Studien,* XXI (1934), 80-103; 125-150.

——, " Der unvollendete Zustand der aristotelisch-scholastischen Metaphysik," *Franz. Studien,* XXIII (1936), 404-425.

——, " Seiender oder werdender Gott? Substanzialität oder Aktualität des Urseienden? ", *Theologische Gegenwartsfragen* (herausgegeben von E. Schlund, Regensburg, 1940, J. Habbel), 63-81.

Buckley, George, *The Nature and Unity of Metaphysics,* Washington, D. C., Catholic University of America Press, 1946.

Byles, W. Esdaille, " The Analogy of Being," *The New Scholasticism,* XVI (1942) 331-364.

DeWulf, Maurice, *History of Medieval Philosophy,* trans. E. C. Messenger, New York, Longmans, Green and Co., 1938, II.

Daniels, Augustinus, " Quellenbeiträge und Untersuchungen zur Geschichte der Gottesbeweise im dreizehnten Jahrhundert mit besonderer Berücktsichtigung des Arguments im Proslogion des hl. Anselm," *Beiträge zur Geschichte der Philosophie des Mittelalters,* VIII (Münster, 1909).

Descoqs, Pedro, *Praelectiones Theologiae Naturalis,* Paris, Gabriel Beauchesne et ses Fils, 1935, II.

Fernandez-Garcia, Maria, *Lexicon Scholasticum Philosophico-Theologicum,* Quaracchi, Coll. S. Bonaventurae, 1910.

Fuchs, Johann, *Die Proprietäten des Seins bei Alexander von Hales,* München, Druck der Salesianischen Offizin, 1930.

Gilson, Étienne, " Avicenne et le point de départ de Scot," *Archives d'Histoire Doctrinale et Littéraire du Moyen Age,* II (1927) 89-149.

——, " Les seize premiers Theoremata et la pensée de Duns Scot," *ibid.* XII-XIII (1937-1938) 5-86.

——, " Les sources gréco-arabes de l'augustinisme avicennisant " *ibid.* IV (1929) 5-149.

——, *God and Philosophy,* New Haven, Yale University Press, 1941.

——, *Réalisme thomiste et critique de la connaissance,* Paris, J. Vrin, 1939.

———, " Metaphysik und Theologie nach Duns Skotus," *Franziskanische Studien,* XXII (1935), 209-231.

———, *The Philosophy of St. Bonaventure,* trans. I. Trethowan and F. J. Sheed, London, Sheed and Ward, 1940.

——— and Philotheus Boehner, *Geschichte der christlichen Philosophie,* Paderborn, Ferdinand Schöningh, 1937.

Grabmann, M., " Die echten Schriften der hl. Thomas v. Aq." *Beiträge zur Geschichte der Phil. des Mittelalters,* XXII (Münster, 1920).

Grajewski, Maurice, *The Formal Distinction of Duns Scotus,* Washington, D. C., Catholic University of America Press, 1944.

———, " Duns Scotus in the Light of Modern Research " in *Proceedings of the American Catholic Philosophical Association,* XVIII (1942) 168-185.

Gredt, Josephus, *Elementa Philosophiae Aristotelico-Thomisticae,* Friburgi, Herder, 1937, I.

Grunwald, Georg, " Geschichte der Gottesbeweise im Mittelalter bis zum Ausgang der Hochscholastik," *Beiträge zur Geschichte der Phil. des Mittelalters,* VI (Münster, 1907).

Harris, C. R. S., *Duns Scotus,* Oxford, Clarendon Press, 1927.

Heiser, Basil, " The Metaphysics of Duns Scotus," in *Franciscan Studies,* XXIII (1942) 379-395.

———, " The Primum Cognitum according to Scotus," *ibid.,* 193-216.

Henry of Ghent, *Summa,* ed. by Hieronymus Scarparius, 3 vols., Ferrariae, apud Franciscum Succium, 1646.

———, *Disputationes Quodlibeticae,* 2 vols., Parisiis, J. Badius Ascensius, 1518.

Hirschberger, Johannes, " Omne ens est bonum," *Philosophisches Jahrbuch,* LIII (1940) 292-305.

Jaeger, Werner, *Aristotle,* trans. Richard Robinson, Oxford, Clarendon Press, 1934.

Jansen, Bernard, " Beiträge zür geschichtlichen Entwicklung der Distinctio Formalis " *Zeitschrift für katholische Theologie,* LIII (1929) 317-344; 517-44.

Klug, H., " L'activité intellectuelle de l'âme selon le B. Duns Scot," *Etudes Franciscaines,* XLI (1929) 517-520.

Knittermeyer, Hinrich, *Der Terminus transszendental in seiner historischen Entwickelung bis zu Kant,* Marburg, Johannes Hamel, 1920.

Kraus, Johannes, *Die Lehre des Johannes Duns Skotus O.F.M. von der Natura Communis,* Freiburg, Studia Friburgensia, 1927.

Kuehle, Heinrich, " Die Lehre Alberts des Grossen von den Transzendentalien " in *Philosophia Perennis,* Regensburg, 1930 I, 129-157.

Longpré, Ephraem, " The Psychology of Duns Scotus and its Modernity," *Franciscan Educational Conference,* XIII (1931) 15-77.

McCall, Raymond, " St. Thomas on Ontological Truth," *The New Scholasticism,* XII, 1938, 9-29.

Maritain, Jacques, *A Preface to Metaphysics,* New York, Sheed and Ward, 1943.

Marston, Robert, *Questiones Disputatae de Anima,* Quaracchi, Coll. S. Bonaventurae, 1932.

Masnovo, Amato, *Problemi di Metafisica e di Criteriologia,* Milan, Società Editrice " Vita e Pensiero ", 1930.

Mercier, Card., *Metaphysique Générale,* Louvain, Institut Supérieur de Philosophie, 1910.

Messner, Reinhold, *Schauendes und begriffliches Erkennen nach Duns Skotus,* Freiburg, Herder, 1942.

Migne, J. P., *Patrologiae Cursus Completus,* Series Latina, 222 vol., Parisiis 1844-1855, (PL.).

Minges, Parthenius, " Der angebliche exzessive Realismus des Duns Scotus," *Beiträge zur Geschichte der Philosophie des Mittelalters,* VII (Münster, 1908).

Ockham, William, *Ordinatio,* revised text edited by P. Boehner (in part printed) *Quaestio Prima prologi* (Schoningh, Paderborn, 1940), remainder in manuscript.

Paulus, Jean, *Henri de Gand* (Études de Philosophie Médiévale, XXV), Paris, J. Vrin, 1938.

Pelster, Franz, " Handschriftliches zur Ueberliererung der Quaestiones super libros Metaphysicorum und der Collationes des Duns Scotus," in *Philosophisches Jahrbuch,* XLIII (1930), 749-787; XLIV (1931) 79-92.

Plassmann, H. C., *Die Schule des hl. Thomas v. Aquino,* Soest, Verlag der Nasse'schen Buchhandlung, 1858, V.

Pouillon, Henri, " Le premier traité des propriétés transcendentales. La Summa de bono du Chancelier Philippe," in *Revue neóscolastique de Philosophie,* XLII (1939) 40-77.

Prantl, Carl, *Geschichte der Logik im Abendlande,* Leipzig, Gustav Fock, 1927, III.

Ross, W. D., " Introduction " to *Aristotle's Metaphysics,* Oxford, Clarendon Press, 1924.

Roth, Bartholomaeus, *Franz von Mayronis, sein Leben, seine Werke, seine Lehre vom Formalunterscheid in Gott,* (Franziskanische Forschungen, hft, 3) Werl in Westfalen, Franziskus Druckerei, 1936.

Schulemann, Günther, *Die Lehre von den Transcendentalien in der scholastischen* (Forschungen zur Geschichte der Philosophie und der Pädagogik, Bd. IV, hft. 2), Leipzig, Felix Meiner, 1929.

Schwendinger, Fidelis, " Zu Binkowski's Arbeiten über die Wertlehre des D. Skotus," *Wissenschaft und Weisheit,* IV (1937) 284-288.

Shircel, Cyril L., *Univocity of the Concept of Being in the Philosophy of John Duns Scotus,* Washington, D. C., Catholic University of America Press, 1942.

Smith, V. E., " On the Being of Metaphysics," *The New Scholasticism,* XVI (1942), 72-84.

Swiezawski, Stephan, "Les intentions premières et les intentions secondes chez Jean Duns Scot," *Archives d'Histoire Doctrinale et Littéraire du Moyen Age,* IX (1934) 205-260.

Thomas of Aquin, St., *Opera Omnia,* ed. by Vivès, 34 voles, Parisiis, 1872-1880.

——, *Summa theologica,* Ottawa, Institute of Medieval Studies, 1941-1944.

——, *Quaestiones Disputatae et Quaestiones Duodecim Quodlibetales,* 5 ed. Taurini, Romae, Marietti, 1927.

——, *De Natura Generis,* ed. Michael de Maria in *Sancti Thomae Aq. Opuscula Philosophica et Theologica,* Tiferni Tiberini, S. Lapi, 1886, I.

Toohey, J. J., "The Term 'Being'," *The New Scholasticism* XVI (1942) 107-129.

Varesius, Carolus Franciscus, *Promptuarium Scoticum,* 2 vols., Venetiis, Andrea Poleti, 1690.

Wilpert, Paul, "Zum aristotelischen Wahrheitsbegriff," *Philosophisches Jahrbuch,* LIII (1940), 3-16.

INDEX

THE CATHOLIC UNIVERSITY OF AMERICA

PHILOSOPHICAL STUDIES

THE SCHOOL OF PHILOSOPHY

---

1. Agnosticism and Religion—An Analysis of Spencer's Religion of the Unknowable. By George Lucas, Ph.D. 1895.
2. Responsibility to Moral Life. By Maurice Joseph O'Connor, Ph.D. 1903.
3. The Knowableness of God. By Matthew Schumacher, C.S.C., Ph.D. 1905.
4. Theory of Physical Dispositions. By Charles A. Dubray, S.M., Ph.D. 1905.
5. The Problem of Evil. By Cornelius Hagerty, C.S.C., Ph.D. 1911.
6. The Basis of Realism. By William Francis Cunningham, C.S.C., Ph.D. 1912.
7. A History of the Theory of Sensation from St. Augustine to St. Thomas. By Othmar Frederick Knapke, C.PP.S., Ph.D. 1915.
8. Classification of Desires in St. Thomas and in Modern Sociology. By Ignatius Smith, O.P., Ph.D. 1915.
9. The Critical Principles of Orestes A. Brownson. By Virgil G. Michel, O.S.B., Ph.D. 1918.
10. St. Thomas' Political Doctrine and Democracy. By Edward F. Murphy, S.S.J., Ph.D. 1921.
11. The Concept of the Human Soul. By William Patrick O'Connor, Ph.D. 1921.
12. The Political Philosophy of Dante Alighieri. By John Joseph Rolbiecki, Ph.D. 1922.
13. Moral Qualities and Intelligence According to St. Thomas. By Joseph Earl Bender, Ph.D. 1924.
14. The Theory of Abnormal Cognitive Processes. By Edward Brennan, O.P., Ph.D. 1925.
15. St. Thomas' Theory of Rationes Seminales. By Michael John McKeough, O.Praem., Ph.D. 1926.
16. Thomism and the New Aesthetic. By Leonard Callahan, O.P., Ph.D. 1927.
17. Aristotelianism in Thomas Aquinas. By Donald Thomas Mullane, Ph.D. 1929.
18. St. Thomas' Theory of Moral Values. By Leo Richard Ward, C.S.C., Ph.D. 1929.
19. Modern Notions of Faith. By Joachim Bauer, O.P., Ph.D. 1930.
20. The Thomistic Theory of Mental Faculties. By Charles A. Hart, Ph.D. 1930.
21. The Humanism of Irving Babbitt. By Francis E. McMahon, Ph.D. 1931.
22. The Substantial Composition of Man According to St. Bonaventure. By Conrad J. O'Leary, O.F.M., Ph.D. 1931.
23. The Phantasm According to the Teaching of St. Thomas. By Sister Mary Anastasia Coady, C.S.N., Ph.D. 1932.

24. Knowledge and Object. By Edward F. Talbot, O.M.I., Ph.D. 1932.
25. Modern War and Basic Ethics. By John K. Ryan, Ph.D. 1933.
26. Man in the New Humanism. By Sister Mary Vincent Killeen, O.P., Ph.D. 1934.
27. Some Modern Non-Intellectual Approaches to God. By Sister Agnes Theresa McAuliffe, C.S.N., Ph.D. 1934.
28. The Order of Nature in the Philosophy of St. Thomas Aquinas. By Joseph Marling, C.PP.S., Ph.D. 1934.
29. The Psychology of St. Albert the Great. By George C. Reilly, O.P., Ph.D. 1934.
30. Cardinal Newman: His Theory of Knowledge. By John Francis Cronin, S.S., Ph.D. 1935.
31. The Primacy of Metaphysics. By Joseph Thomas Casey, Ph.D. 1936.
32. The Philosophy of Athenagoras: Its Sources and Value. By Henry Albert Lucks, C.PP.S., Ph.D. 1936.
33. The Philosophical Basis for Individual Differences According to St. Thomas Aquinas. By Robert Joseph Slavin, O.P., Ph.D. 1936.
34. Lawlessness, Law, and Sanction. By Miriam Theresa Rooney, LL.B., A.B., A.M., Ph.D. 1937.
35. The Substance Theory of Mind and Contemporary Functionalism. By Thomas J. Ragusa, Ph.D. 1937.
36. The Psychology of St. Bonaventure and St. Thomas Aquinas. By Clement M. O'Donnell, Ph.D. 1937.
37. The Rational Nature of Man. By James C. Linehan, S.S., Ph.D. 1937.
38. The Metaphysical Foundations of Dialectical Materialism. By Charles J. McFadden, O.S.A., Ph.D. 1938.
39. The Problem of Solidarism in St. Thomas. By Sister Mary Joan of Arc Wolfe, Ph.D. 1938.
40. St. Thomas' Doctrine of Substantial Form, and the Relations Between This Doctrine and Certain Problems and Movements of Contemporary Philosophy. By Brother Benignus Gerrity, F.S.C., Ph.D. 1937.
41. The Problems of Altruism in the Philosophy of Saint Thomas. By Cyril Harry Miron, O.Praem., Ph.D. 1939.
42. De Occultis Operationibus Naturae, According to St. Thomas Aquinas. By Joseph B. McAllister, S.S., Ph.D. 1939.
43. The Philosophy of Personality in the Thomistic Synthesis and in Contemporary Non-Scholastic Thought. By James H. Hoban, Ph.D. 1939.
44. Potentiality and Energy. By Edward M. O'Connor, Ph.D. 1939.
45. The Absolute and the Relative in St. Thomas and in Modern Philosophy. By Sister Mary Camilla Cahill, C.D.P., Ph.D. 1939.
46. The Antecedents of Being: An Analysis of the Concept 'De Nihilo' in St. Thomas' Philosophy. By Sister Mary Consilia O'Brien, O.P., Ph.D. 1939.
47. Punishment, in the Philosophy of St. Thomas Aquinas, and Among Some Primitive Peoples. By George Quentin Friel, O.P., Ph.D. 1939.
48. The Social Value of Property According to St. Thomas Aquinas. By William J. McDonald, Ph.D. 1939.
49. The Philosophy of Labor in St. Thomas. By Sylvester Michael Killeen, O.Praem., Ph.D. 1939.
50. The Family. A Thomistic Study in Social Philosophy. By Anthony Leo Ostheimer, Ph.D. 1939.

51. FIRST PRINCIPLES IN THOUGHT AND BEING. By James Bacon Sullivan, Ph.D. 1939.
52 THE THEORY OF KNOWLEDGE OF SAINT BONAVENTURE. By Sister Mary Rachel Dady, O.S.F., Ph.D. 1939.
53. THE METAPHYSICAL FOUNDATIONS OF THOMISTIC JURISPRUDENCE. By Karl Kreilkamp, Ph.D. 1939.
54. THE CONCEPT OF SELF IN BRITISH AND AMERICAN IDEALISM. By Hugh Joseph Tallon, Ph.D. 1939.
55. THE SEPARATED SOUL IN THE PHILOSOPHY OF ST. THOMAS AQUINAS. By Victor Edmund Sieva, Ph.D. 1940.
56. EFFICIENT CAUSALITY IN ARISTOTLE AND IN ST. THOMAS. By Francis X. Meehan, Ph.D. 1940.
57. CHARACTER CONTROL OF WEALTH ACCORDING TO ST. THOMAS AQUINAS. By Sister Francis Augustine Richey, Ph.D. 1940.
58. ESSENCE AND OPERATION IN THOMISTIC AND IN MODERN PHILOSOPHY. By Sister Mary Dominica Mullen, R.S.M., Ph.D. 1941.
59. THE INTELLECTUAL VIRTUES ACCORDING TO THE PHILOSOPHY OF ST. THOMAS AQUINAS. By Sister Rose Emmanuelia Brennan, S.H.N., Ph.D. 1941.
60. UTOPIAS AND THE PHILOSOPHY OF ST. THOMAS AQUINAS. By Sister M. St. Ida Le Clair, P. de M., Ph.D. 1941.
61. A PHILOSOPHY AND PSYCHOLOGY OF SENSATION ACCORDING TO ST. THOMAS AQUINAS. By Jerome Paul Ledvina, Ph.D. 1941.
62. A THEORY OF CRITICISM OF FICTION IN ITS MORAL ASPECTS ACCORDING TO THOMISTIC PRINCIPLES. By Sister Mary Gonzaga Udell, O.P., Ph.D. 1941.
63. THE MORAL BASIS OF SOCIAL ORDER ACCORDING TO ST. THOMAS. By George V. Dougherty, Ph.D. 1941.
64. LIBERTY UNDER HISTORICAL LIBERALISM AND TOTALITARIANISM. By Rudolf Harvey, O.F.M., Ph.D. 1941.
65. THE QUALITIES OF CITIZENSHIP IN ST. THOMAS. By Gerard R. Joubert, O.P., Ph.D. 1941.
66. A CRITIQUE OF THE PHILOSOPHY OF GEORGE SANTAYANA IN THE LIGHT OF THOMISTIC PRINCIPLES. By Sister Cyril Edwin Kinney, O.P., Ph.D. 1942.
67. THE UNIVOCITY OF THE CONCEPT OF BEING IN THE PHILOSOPHY OF JOHN DUNS SCOTUS. By Cyril L. Shircel, O.F.M., Ph.D. 1942.
68. THE PHILOSOPHY OF EQUALITY IN THE PHILOSOPHY OF ST. THOMAS AQUINAS. By Sister Jane Frances Ferguson, S.M., Ph.D. 1942.
69. VARIOUS GROUP MIND THEORIES VIEWED IN THE LIGHT OF THOMISTIC PRINCIPLES. By Sister Mary Dolores Hayes, R.D.C., Ph.D. 1942.
70. THE THOMISTIC CONCEPTION OF AN INTERNATIONAL SOCIETY. By Gerald Francis Benkert, O.S.B., Ph.D. 1942.
71. THE SOCIAL VALUE OF HOPE IN MODERN SOCIAL THOUGHT AND THOMAS AQUINAS. By Clement Della Penta, Ph.D. 1942.
72. THE ACT OF SOCIAL JUSTICE IN THE PHILOSOPHY OF ST. THOMAS AQUINAS AND IN THE ENCYCLICALS OF POPE PIUS XI. By William Ferree, Ph.D. 1942.
73. A THOMISTIC ANALYSIS OF SOCIAL ORDER. By John F. Cox, Ph.D. 1943.
74. THE PHILOSOPHY OF MODERN REVOLUTION. By James J. Maguire, C.S.P., Ph.D. 1943.
75. THE NATURE OF DEMONSTRATIVE PROOF ACCORDING TO THE PRINCIPLES OF ARISTOTLE AND ST. THOMAS AQUINAS. By Owen Bennett, O.M.C., Ph.D.. 1943.

76. The Social Role of Truth According to St. Thomas Aquinas. By Alan J. McSweeney, C.P., Ph.D. 1943.
77. Selfishness and the Social Order. By John J. Reardon, C.P., Ph.D. 1943.
78. Social Leadership According to Thomistic Principles. By Luke Fisher, S.A., Ph.D. 1943.
79. The Social Role of Self-Control. By James Kerins, C.SS.R., Ph.D. 1943.
80. The Totalitarian Philosophy of Education. By Leonard Bernard Pousson, Ph.D. 1944.
81. The Philosophical Theory of Creation in the Writing of St. Augustine. By Christopher J. O'Toole, C.S.C., Ph.D. 1944.
82. Psychology and Philosophy of Teaching According to Traditional Philosophy and Modern Trends. By Alfonso Vargas, F.S.C., Ph.D. 1944.
83. A Philosophy of Poetry Based on Thomistic Principles. By John A. Duffy, C.SS.R., Ph.D. 1944.
84. The Theory of Natural Appetency in the Philosophy of Saint Thomas Aquinas. By Gustaf J. Gustafson, S.S., M.A., Ph.L. 1944.
85. The Importance of Rural Life According to the Philosophy of St. Thomas Aquinas. By George H. Speltz, Ph.D. 1944.
86. Metaphysics as a Principle of Order in the University Curriculum. By Alfred F. Horrigan, Ph.D. 1944.
87. The Theory of Knowledge of Hugh of St. Victor. By John P. Kleinz, Ph.D. 1944.
88. The Categories of Being in Aristotle and St. Thomas. By Sister Marina Scheu, Ph.D. 1944.
89. The Thomistic Philosophy of the Angels. By James D. Collins, Ph.D. 1944.
90. The Formal Distinction of Duns Scotus. By Rev. Maurice J. Grajewski, O.F.M., Ph.D. 1944.
91. Philosophy of Social Change According to the Principles of St. Thomas. By Brother E. Stanislaus Duzy, Ph.D. 1944.
92. The Intelligibility of the Universe in the Philosophy of St. Thomas Aquinas. By Sister Mary Cosmas Hughes, R.S.M., Ph.D. 1946.
93. The Social Thought of St. Bonaventure, the Seraphic Doctor. By Reverend Matthew De Benedictis, O.F.M., Ph.D. 1946.
94. The Nature and Unity of Metaphysics. By Reverend George Buckley, M.M., Ph.D. 1946.
95. A Critique of the Philosophy of Being of Alfred North Whitehead in the Light of Thomistic Philosophy. By Reverend Leo Foley, S.M., Ph.D. 1946.
96. The Transcendentals and their Function in the Metaphysics of Duns Scotus. By Reverend Allan B. Wolter, O.F.M., Ph.D. 1946.

Copies may be obtained at

The School of Philosophy

The Catholic University of America

Washington, D. C.